# Liberation and Technology

## Development possibilities in pursuing technological autonomy

Gussai Sheikheldin

**MKUKI NA NYOTA**

DAR—ES—SALAAM

PUBLISHED BY
Mkuki na Nyota Publishers Ltd
P. O. Box 4246
Dar es Salaam, Tanzania
www.mkukinanyota.com

© Gussai Sheikheldin, 2018

ISBN 978-9987-08-329 9

Visit www.mkukinanyota.com to read more about and to purchase any of Mkuki na Nyota books. You will also find featured authors, interviews and news about other publisher/author events. Sign up for our e-newsletters for updates on new releases and other announcements.

Distributed world wide outside Africa by African Books Collective.
www.africanbookscollective.com

# Contents

*To my compass, M. M. Taha,*
*to Amilcar Cabral,*
*to Ursula Franklin,*
*and to those who acknowledge the decisive role of praxis*

# Acknowledgements

This work came to fruition through a process of learning, digesting, contemplating, synthesis, articulation, and editing. One could say this is customary for any work of this genre. Through each phase the author(s) consults and coordinates many sources, to the extent that no final product could be described as the authentic brainchild of the author(s) alone, in a proper sense. That said, of course, the intellectual responsibility of the final product rests with the author(s), since the arguments were devised and expressed by them and rely in the end on their judgment. Acknowledgements are a humble attempt to give credit to the main entities to whom the author(s) owes a debt for their contribution to the work. Here is my imperfect, concise attempt.

I would like to begin with acknowledging some of the intellectuals, scholars, and trailblazers I learned from and consulted amply for this work: Mahmoud M. Taha, Karl Polanyi, Julius K. Nyerere, Lewis Mumford, Amilcar Cabral, Elinor Ostrom, Johan Galtung, Abdullahi An-Na'im, Walter Rodney, Ursula Franklin, John Garang de Mabior, M. Jalal Hashim, Issa Shivji, Ali Mazrui, and others.

I thank Bitrina Diyamett (and the Science, Technology and Innovation Policy Research Organization – STIPRO) for the relevant intellectual exchanges and valuable comments on earlier drafts of this work. I thank Livinus Manyanga (and KAKUTE) for the very valuable experience of engaging in practice with technological development work, which is undertaken with vision and resilience, under difficult constraints in developing communities and their surrounding environments.

I thank John Devlin, Nonita Yap, and Gail Krantzberg, for guiding me through a long and disciplined journey of cultivating scholarly knowledge and understanding of topics relevant to this work.

The chapters of the book went through multiple reviews and drafts. I thank all those who participated in that process with their time, insight and skill.

And, of course, there is much to be said about the stimulation, encouragement, and support we receive from our human interactions in our lifeworlds; particularly those that indirectly provide us with drive, facilitation, and tacit knowledge of what we are set to do. I certainly am indebted to many family members, friends, partners and colleagues in that respect. I will not attempt to list them by name here, however. They mostly know who they are.

I hope that this work reflects the contributions of all those mentioned in a good light. I take responsibility for any failure in that respect.

# Abbreviations

| | |
|---|---|
| CNC | computer numerical control |
| DFI | direct foreign investment |
| GDP | gross domestic product |
| GII | Global Innovation Index |
| ICT | information and communications technology |
| ITK | indigenous technical knowledge |
| NIE | Newly Industrializing Economy |
| NIS | national innovation system |
| NTC | national technological capabilities |
| OECD | Organization of Economic Co-operation and Development |
| R&D | research and development |
| S&T | science and technology |
| STI | science, technology and innovation |
| TAI | Technology Achievement Index |
| T&S | technology and science |
| UNCBD | United Nations Convention on Biological Diversity |
| UNDP | United Nations Development Programme |
| WCD | World Commission on Dams |
| WTO | World Trade Organisation |

# Introduction

*"Technology has built the house in which we all live."*

–Ursula Franklin[1]

In the documentary series, *The Africans: A triple heritage*, prepared and hosted by the late Ali Mazrui, in one part he took the viewers through big, state-of-the-art industrial factories in two West African countries. Those factories were built right after political independence was gained, and they were built from the ground-up paid for by their respective states. They were built to represent a proud leap forward by these newly independent countries towards the era of modern industrialization. However, people and products were missing from the floors of those factories. Mazrui walked through these empty factory floors and told their story. They were decommissioned plants that lasted for only short periods of time before the governments realized they were running at a cost higher than their return, with no change foreseen in that situation in the near future. These were high-tech plants, not only commissioned and built by western corporations, but also operated by them on contracts with the government. The plants were foreign plants on African soil, and they were unable to serve the developmental priorities of their host countries. In the end a tough lesson was learned (hopefully) the hard way.

These days, decades after political independence, it is not a rare sight in any developing country, in urban areas, to find most, if not all, the high-tech products the world offers today, whether in stores and along streets, or in buildings, hotels, and factories. You find new cars

---

1    In *The Real World of Technology.* CBC Massey lectures, November 1989.

on tarmacked roads, modern electronics and other products in these supposedly technologically-challenged societies. It is not the absence of advanced technological products in their markets, and some of their industries, that is the problem, but rather the matter of quality and how ubiquitous such products are. Quality is often compromised in that most of the contemporary technological products found in developing societies are imported and not subject to sufficient regulatory scrutiny. Less commonly, but to the same effect, products are designed somewhere else – especially the complex parts – and assembled locally due to this weak regulatory environment as well as lower labour costs and taxes. Such products are rarely local products in any real sense. There is also no full integration of these contemporary technologies into the lives and work of the majority of the population. The limited quantity of these technological products and services sees the majority of the population living in conditions of absence of essential products and services that the populations of industrialized countries enjoy and might take for granted. Those essential products and services include access to electricity, clean water, civil infrastructure, contemporary healthcare systems and aiding machinery, as well as formal education.

The modern world is a spectrum characterised by two extremes of which one part is lacking the basics of a modern life aided by contemporary technological systems, while the other part, on the other hand, is what can be called hyper-technologized, or ultra-technologized. Between these two extremes, there are other parts where there is a presence of industrialization and material progress that is not inclusive of the entire area, and parts with fairly industrialized societies. The issue is not materialist in essence, however. It is not about whether having more contemporary technology is a good or a bad thing in itself. The issue is rather about what each society's relationship with technology today says about the quality of the lives of the people of that society.

Processes of technological change pose many challenges for developing societies. They tend to be complex and multi-faceted, involving numerous variables, agents, and contexts. However, they are a critical part of economic and human development for all societies. Historical evidence shows a strong correlation between technological development and human development (Hill and Dhanda 2003). The United Nations Development Programme (UNDP) shows how seminal advances in human development in the 20th century were largely attributable to technological improvements and breakthroughs, in different sectors (e.g. health and hygiene, agriculture, transportation, etc.) (UNDP 2001), with the term 'industrialized economies' coming to be widely used to refer to advanced, wealthy countries where these technological developments occurred.

Development in different sectors requires different stimulants and interventions depending on the social and cultural context. For example, in Africa, where agriculture is historically of great importance, slow rates of adoption of new technologies and minimal increases in productivity are the norm. Researchers have pointed to subjective and social challenges facing agricultural technology adoption in Africa, such as farmers' negative perceptions of technological changes, or cultural barriers to accepting them.[2] Dercon and Christiaensen (2011) demonstrate that besides a more general subjective resistance farmers consider crude cost-benefit analyses and the multitude of household priorities, which lead many of them to avoid adopting new technologies. The story of Africa and agriculture resonates, in varying degrees, with other developing regions such as Southeast Asia and Latin America.[3]

In contrast, information and communication technologies (ICTs) have in a relatively short period experienced rapid adoption in a variety of other sectors in developing societies around the world. ICTs have been adopted in healthcare, tourism, small and medium-sized enterprises, and in education.[4] Between the two extremes of a lack of new technology adoption in agriculture and the fast uptake of ICTs in a variety of sectors, there are varying degrees of technology adoption within other sectors in developing societies, such as in water and sanitation, alternative energy, and small industries.

Moreover, processes of technological change and industrialization across societies show varying patterns and degrees of success. Measures of technological innovation capacity and output such as the Global Innovation Index (GII) and the Technology Achievement Index (TAI)[5] suggest that there is no roadmap appropriate for all countries pursuing technological progression. The historical paths of industrialization followed by the USA, the UK, Germany, Russia, China, Japan, India

2    Stamp 1990; Simalenga 1999; Adesina 1995; Rauniyar 1992.
3    Binswanger 1986; Adeel et al. 2008; Martinez-Torres et al. 2010.
4    UNDP 2001; Lekoko and Semali 2012; Rensburg et al. 2008; Nasir et al. 2011.
5    The TAI is used by the UNDP to measure a country's technological capacity and progress in comparison to other countries. The TAI uses four dimensions of technological capacity: creation of new technology; diffusion of recent innovations; diffusion of old innovations that are still fundamental for industrialization; and the building of a human skills base for technology origination and adoption. Each one of the dimensions has two statistical sub-indicators (Desai et al. 2002).

and Brazil are dissimilar and largely influenced by local variables, including factor endowments, socioeconomic institutions, market relations, policies and sociopolitical histories. In addition, these indices show that some developing regions have made almost no technological progress over long periods of time, leaving a huge gap between them and countries that have progressed (Desai et al. 2002).

In spite of this gap between technologically advanced and technologically challenged societies, the indices suggest that technological change is building up globally. Despite serious setbacks in some contexts, technological knowledge and skills of local populations and global interconnectedness of technological markets, procedures and research and development (R&D) methods are increasing overall (Desai et al. 2002; Nasir et al. 2011).

There are good reasons why most technological capacity assessments focus on national scales, since indicators at national levels rely on more accessible data inputs. Yet it also makes sense to assess technological capacity (and achievements) with smaller and more 'organic' social aggregates – i.e. aggregates formed around ecological or socioeconomic relations such as communities of geo-ecological regions and industrial clusters. Some attempts at explaining this phenomenon refer to evidence that such social aggregates tend to correlate with 'technological hubs' within countries. According to the Global Development Index Report, these hubs – such as Silicon Valley-type industrial clusters – are usually responsible for painting the entire country's technology mode or level, suggesting that the mode is evenly distributed across the whole country when it is actually concentrated in a few places (Dutta and Lanvin 2013). For example, the Silicon Valley industries make the state of California, and indeed the whole USA, appear as a global leader in ICT technology, when in fact it is really the Silicon Valley cluster that is the global ICT leader.

Different indices confirm that there are indications of a statistically significant, positive correlation between technological achievement and human development—this is apparent in a comparison between the UNDP-initiated indices: Technology Achievement Index and Human Development Index (Hill and Dhanda 2003, 29). The technology divide between countries of the world appears to be a strong indicator of the human development divide as well.

From this broad overview, several general positions take shape: 1) that technological change is important for development, 2) that it evolves in multiple ways; 3) that it can be measured in a variety of ways; 4) that there may be alternative ways to explain the main features of its evolution and diversity; and 5) that sustained economic development requires increasing local capacity to use, control and maintain technosocial systems. Such technosocial systems refer to people and technologies working in combined efforts that form functional wholes (Woodhouse and Patton 2004).

Discussions continue in international development circles about the importance of developing and increasing endogenous technological capacities (see for example Shaw 2002; Adeel, Schuster and Bigas 2008; Nasir et al. 2011). The discussions are not so much about whether endogenous capacities are important, but about what levels of such capacity are needed to advance the economic and human development agenda. This book takes the position that higher levels of endogenous technological capacity are necessary to achieve development in key sectors, such as science and education, agriculture, energy, water supply, health and hygiene, infrastructure and basic industries.[6]

## Thesis

The main thesis of this book is that if developing societies seek genuine human and socioeconomic development then they need to seek technological autonomy. Technological autonomy refers to the attainment of a sufficient level of self-determination in generating and managing technological phenomena for that society. It means acquiring an endogenous capacity to generate, transfer and administer technologies, as well to guide policies and manage innovation, industrial sectors, local and foreign trade, and priorities of development. Such autonomy also, of course, implies a relative independence from external manipulation; particularly from other societies with greater economic, political and military power. Technological autonomy, therefore, is a concept that identifies a policy and sociopolitical approach, with key consideration for technical and economic factors, to the issue of technological change and development. Consequently, it is an approach that involves institutional as well as technological affairs.

On a broader perspective, there exists a consensus that technology has an omnipresent power in all contemporary societies. Whether in developing or developed contexts, the power of technology in contemporary lives everywhere is a power with measurable and penetrating authority—like a fifth estate.

---

6    Mazrui 1986; Haug 1992; Nyerere 1968 and 2011; STIPRO 2010; Page 2016

This claim is not made lightly. The terminology of a fourth, fifth, etc. estate came to use to refer to a phenomenon where a particular sector in modern society has a strong and observable influence in that society but without the direct allocated powers the 'first' three estates – or branches of government: the executive, legislative and judicial – possess. Obviously, the notion of additional estates (or authorities) assumes the presence of state systems that run along the doctrine of the separation of powers, a condition not applicable to all societies today. Nonetheless the term has taken on a life of its own, unbound by the context from which it sprang. Today everywhere in the world we can say 'the fourth estate' and be understood as reference to the press and media. Recently, there has been a competition over the title of the fifth estate, with some crediting it to the new independent and global cyber media (e.g. bloggers, e-journalists, hacktivists, and non-mainstream media outlets) which is more decentralized than conventional media and press in terms of control, and more influential, some argue, than these traditional institutions. I would argue that these new media forms certainly bring something qualitatively new to the table, but not to the point of establishing a fifth estate separate from the fourth. These new media forms represent a revolution within the fourth estate itself, a revolution that should actually be credited largely to technology; particularly ICT.

Rather, I would argue that it is more sensible, and high time, that the omnipresence and influence of technology in our lives today be recognized and addressed in ways more cognizant of that reality. We know that "it is largely by technology that contemporary society hangs together" (Franssen et al. 2013), so why should we not be explicit? The fifth estate is technology. By that we mean that the various decentralized and interrelated institutions and agents that create and regulate technology in our societies operate and define the apparatus of the fifth estate. These institutions and agents have visible, wide and deep power, albeit 'unofficial', in shaping our lives.

Guided by this broad perspective on the place of technology in society, this book addresses the implications this has in the context of technological change in developing societies.

The existing literature on technological change is relatively diverse and abundant, yet not sufficiently integrated. There is considerable scholarship on the theory and historical analysis of developing technological capabilities in firms and national systems.[7] There is work on technological change models, as they relate to dynamics of markets, resources and stimulation of industrial innovation.[8] Further, there are established fields that relate to technological change and overlap with it, such as diffusion of innovations[9] and institutional economics of technology affairs.[10] There is also an influential literature on understanding the phenomenon of technology as it pertains to both developed and developing societies.[11] As for developing societies there is extant literature on the dichotomies between traditional and modern technologies[12] as well as the role of national and international dynamics in technology transfer and technological development as it relates to various factors, such as development policies and international relations.[13]

Few works, however, connect the multitude of themes mentioned above through conceptual frameworks that integrate and map big pictures. One such framework is the National Innovation System (NIS) framework, which aims to organize productive forces and structures, and the flow of information and skills in a country, in order to increase the output of innovative solutions to development constraints (Maharajh, Scerri and Sibanda 2013). In that framework, science, technology and innovation (STI) play a central role, and thus require strategic investment. At the policy level the NIS will include careful investments in education systems, enterprise support and labour markets (Lundvall 1992). Many countries are careful to devise and improve their own NISs as part of their national development plans. This framework operates only at the national policy level by default and thus contributes to the design of macroeconomic policies. It is, however, vague on key technological activities that are not considered 'innovative', but customary or traditional, even if they are recognised as important for the particular context. There are also a few other frameworks with limited use and scope (for example, see Aubert 2005).

The framework of technological autonomy, presented in this book, builds on the literature discussed above and ameliorates some of the

7   Lall 1992; Wolff 1999; Kim and Nelson 2000; Oyelaran-Oyeyinka& McCormick 2007; Mazzucato 2013.
8   Dosi 1982; Arthur 1989; Ruttan 1997.
9   Wejnert 2002; Rogers 2003; Huh and Kim 2008; Haider &Kreps 2010; Zanello et al. 2015.
10  Polanyi 1944; Rosenberg 1982; Binswanger 1986; Kroszner 1987; Haug 1992; Page 2016.
11  Mumford 1967 and 1970; Galtung 1979; Franklin 1989; Aunger 2010; Franssen et al. 2013.
12  Hyman 1987; Gamser 1988; Scott 1999; Roy 2002; Adeel, Schuster and Bigas 2008.
13  Morehouse 1979; Nyerere 1998; UNDP 2001; Shaw 2002; Diyamett&Risha 2015.

mentioned shortcomings of other conceptual frameworks. It describes processes of technological change in developing societies. If developing societies seek to improve levels of human and economic development it will be necessary for them to develop an endogenous capacity to oversee technological affairs. It is this capacity that we can term "technological autonomy". Such autonomy includes a "strengthened autonomous capacity for creating, acquiring, adapting and using technology" (Morehouse 1979, 387) and an autonomous decision-making capacity to plan and manage the local affairs of industrial and infrastructural development. Further, this framework presents two main variables that lead to technological autonomy: technology localization and technological capabilities. Technology localization consists of three activities: diffusion, institutional support, and technical adaptation. Technological capabilities, on the other hand, consist of: production activities, investment activities and networking of actors who generate technological innovations and knowledge. Together, advances in technology localization and technological capabilities advance a society along the path towards technological autonomy.

The groups, bodies and individuals who actualize and set in motion the process of technological change, undergirding technological autonomy, are called agents of technological change – such as the state, private industries, and non-governmental organizations. They activate and support – i.e. operationalize – the mechanisms of the technological autonomy framework. This proposed framework identifies the main elements of technological change and helps to visualize and connect its goals and objectives in developing societies.

## Liberation Technology

This study has a particular approach to the concept of liberation which sees the concepts of decolonization and autonomy as related and related in turn to technological change in contemporary developing societies.

Decolonization is the process of recuperating from the experience of being colonized. Perceiving decolonization as a process is vital because decolonization thought of as an 'event' then most will associate it with the declaration of political independence and the transfer of political power from a foreign administration to a government of native faces (i.e. different faces in high places). This event is not decolonization, but at most one milestone in the process. Decolonization is bigger, deeper, and more complex, and that is most evident in that, often enough, foreign colonizers get replaced by native oppressors in the experience of the masses. One can say that a genuine decolonization process is complete when genuine autonomy is attained, and vice versa. In its relation to technological development, decolonization would mean breaking away from the colonial relationships of one-directional technology transfer, as well as trade of raw materials exports vs. finished products import, etc. It would also mean a more comprehensive fostering of the work of harnessing native technological capabilities to the point where the local industrial relations and innovations begin to express themselves without being helplessly tied to the politics, markets and technological institutions of former colonizers. It would mean reaching technological autonomy.

Liberation, as a concept used in this book, takes the whole process of decolonization a step further. We can say that decolonization is one phase, or part, of a society's liberation process. While the goal of decolonization is breaking away from the remnants of colonization, the goal of liberation is breaking away from all material conditions that limit and inhibit the population. Liberation is thus the successful effort to minimize (or, preferably, eliminate) dependency and exploitation. The act and process of liberation is transformative and comprehensive, and takes place over time (i.e., is not incidental or momentary). In the context of technological development in developing societies, liberation amounts to harnessing technological capabilities and advances to the point where good and sustained measures of human development are achieved, and where there is freedom from substantial barriers to living a flourishing life; a freedom attained and maintained using an autonomous process of building, learning, growing and choosing.

Liberation in the context of technological development is relatively, reminiscent of Amartya Sen's approach of 'development as freedom' (1999). Sen argues that, in contrast to the narrower views of development (e.g., dry and broad econometrics with generalized 'averages'), development can be seen as the expansion of "substantive human freedoms" to lead the kind of life we value (as human individuals and as human communities). Pursuing such development amounts to eliminating those things that limit political freedom, economic facilities, social opportunities, transparency guarantees, and protective security. Rearticulated: development is the process of creating conditions whereby obstacles are eliminated or minimized, and opportunities are enhanced, for all individuals and groups as they seek to realize their full potentials and aspirations. Liberation is understood along similar lines. So, for this book, while technological change is the main topic, it is perceived as a mean to an end, and that end is liberation.[14]

Yet liberation implies a proactive process rather than a status to be achieved (that is freedom). So here we are concerned with the liberating process of technological change—i.e. a technological change that has liberation as its focus. This approach to "liberation technology", was chosen before the author later on came across a group that used the same term in a legitimate, but more limited, context. The Center for Democracy, Development and Rule of Law at Stanford University uses the term to refer to how information technology can be used to "improve governance, empower the poor, defend human rights, promote economic development, and pursue a variety of other social goods."[15] Here liberation, and the role of technology in it, is conceived in ways that can include all the above and more; but importantly more. Essentially, all technologies can be, as some already are, used to further goals of human development, prosperity and dignity. Technologies of agriculture, energy harvesting and distribution, water and sanitation, healthcare, local value chain development, transportation, communication, etc., all can be engaged as liberation technologies, as described in this book. If they are integrated and utilized to serve the elimination of dependency and exploitation, they are liberation technologies.

The approach of this book to liberation is influenced by two movements: the liberation theology movement and the legacy of the anti-colonial liberation movements worldwide.

The movement of liberation theology took shape in the late 20th century in Latin America, led by proponents of Christian Catholicism. It sought to engage religious and moral discourse into siding with the poor and the oppressed and taking a stand against 'sinful' socioeconomic practices that dispossess and exploit the vulnerable folk in society. It was essentially a sociopolitical movement that was guided by a theological worldview that, as part of its framework, perceived social justice as a moral stance with the goal being to mobilize for alleviation of conditions inimical to the realisation of social justice. "When I give food to the poor,

---

14    It is understood, however, that conventional indicators of development are not totally irrelevant. There is justification for using them when deemed suitable within the larger scope of liberation. Measurement like rise in incomes and purchasing power, industrialization, technological advances and social modernization can have their proper place in the larger picture if used appropriately.

15    Center for Democracy, Development and the Rule of Law, Program on Liberation Technology website page: http://cddrl.fsi.stanford.edu/libtech/ (visited December 16, 2015).

they call me a saint. When I ask why the poor have no food, they call me a communist," said Dom Hélder Pessoa Câmara, one of the prominent figures associated with the liberation theology movement. Another known figure of that movement is South African Archbishop Desmond Tutu, whose record of anti-apartheid struggle was consistent with his words that, "to be neutral in a situation of injustice is to have chosen sides already. It is to support the status quo." An intellectual representative of the liberation theology movement was Paulo Freire, the educator and philosopher who wrote *Pedagogy of the Oppressed* (1984). In his seeking to merge adult/literacy education with sociopolitical critical awareness, Freire introduced to the global critical literature the concept of 'conscientization' which is defined as the process of becoming critically conscious of structural sources of oppression in society as obstacles to genuine development. The process of education, to Freire, is a process of engagement with the masses. "Human existence cannot be silent, nor can it be nourished by false words, but only by true words, with which man transforms the world. To exist, humanly, is to name the world, to change it." (Freire 1984, 77). Freire's work is mainly about promoting education—not any education, but one engaged in critically addressing social reality—as a conscientization and emancipation process. Critical education is thus one field where theory and practice come together (praxis). Similarly, technological development is another field in which the call to praxis – conscious, strategic planning and implementation – is made.

The main lesson taken from liberation theology is that theology by itself can support one position or another in social dynamics. Theology can be, and had been, used to further interests of elites and exploitative trends in history. Yet it can also be used to further progressive notions of liberation, social justice, empathy and mindful action for desired change. Indeed it has been used in that manner multiple times in history as well. The same, in that particular quality, can be said of technology. By itself, technology takes no particular sides in social dynamics. The gears will turn—if assembled correctly—regardless of who is turning them. An automotive vehicle will obey the laws of physics and move from point A to point B if all the conditions for its movement are satisfied. Yet technology can also be deliberately employed and integrated in a process of liberation; as described above and as will be discussed in the book.

# Book outline

Following this introduction, chapter one presents a theoretical elaboration for understanding and perceiving technology, institutional dynamics, and technological change. It explores and builds a definition of technology from a historical and developmental perspective. This exercise in defining technology seeks to clarify technology's function in human existence, particularly the socio-ecological existence. The chapter then discusses the three main conditions that influence technological change processes in developing societies: technology-institutional dynamics, the dichotomy between traditional and modern technology in key sectors, and development priorities of societies. The chapter also demonstrates some models of technological change, and examples of different manifestations.

Chapter two contains the main proposal of this book, which is that in addressing technological change processes in developing societies technological autonomy is paramount. The chapter begins with proposing and explaining the framework and fleshing-out its variables (purpose, tools, and elements). The discussion of technological autonomy shows this to be the attainment of a sufficient level of self-determination in planning and managing technological matters for a society.

Chapter three sheds light on the agents of technological change—the bodies, groups and organizations that initiate and operationalize technological change processes in contemporary societies. The chapter identifies which the chief agents (the state and a few others) are, in addition to their characteristics, what roles they often play and under which circumstances.

Chapter four discusses important sources of influence on technological affairs of societies, such as the political atmospheres, ecological systems and cultures (or cultural institutions). Within each of these influences, and in their interfaces, technology permeates complex institutions and navigates through tough balances. This chapter touches on broad, and seemingly sporadic, subjects, but the general point is that technology interacts extensively with each of the grand influencing phenomena shaping our contemporary lives and we cannot afford reductionism when engaging technological change processes. We live within states, exist under complex ecosystems, and interact through very diverse and persisting cultural and social systems – such as educational systems, languages, division of labour, communication networks, etc. This chapter explores how to navigate through this myriad of elements and connections with technological change in mind.

Chapter five is dedicated to one specific point within the main topic of liberation technology: technology and justice. The choice to dedicate a chapter to the topic of justice is because justice is central to development and to liberation, and this book is about technological development as a vehicle for human development and freedom. Sustainable development seeks that people attain better lives because they are worthy of it by virtue of being human. That is a value judgement on human life— it is the worthiness of humans that entails they should have their needs satisfied, live in reasonable comfort and that there be room for expressing aspirations. Because of that value judgement we find that measures of pure material progress do not suffice in expressing matters of technological development from this perspective. The topics the chapter addresses include issues of modernization vs. westernization, as well as brain drain migration from the economic south to the north (as a problematic phenomenon for developing societies and beneficial for industrialized ones). Issues of gender and technology in developing societies are also addressed. A discussion of alienation and dispossession as they relate to the challenges of technological change is also included. The chapter also talks about ICTs as a double-edged sword that could work either for supporting and enhancing justice or for subverting it (depending on how it is used). The chapter also talks about justice in the factory, which is the temple of modern technology and where it is both optimally utilized and gives birth to the products that shape contemporary lives. Finally, a chapter on technology and justice would not be complete without addressing the issue of poverty in the presence of globally capable technology.

Chapter six, being the last, is relegated to further discussions – or selected stories – that generally take the form of case studies. These discussions and stories are meant to bring together the various issues mentioned in the former chapters and examine them in real historical cases. It first discusses the phenomenon of appropriate technology – what it means, what are its goals and how it came about – followed by a critique of theory and practice relevant to today's international development circles. The second discussion is a look at dams and development, as mega technological projects (large dams) in contexts of active pursuit of energy (hydropower) and agricultural growth (irrigation). Stories of dams and development from two different developing regions of the world, the Nile basin and India, provide substance for a discussion of large dams in developing countries: how much they cost in economic, social and ecological costs and how do these costs measure compared to the alleged benefits. Finally, the chapter discusses a particular story of a countrywide rural development scheme in as far as it has a theoretical affinity to technological self-reliance and localizing development visions—the story of Ujamaa rural development scheme in Tanzania, in the third quarter of the twentieth century. Was the practice, or implementation, of Ujamaa consistent with its original vision? And what can we learn from that unique experience?

A short section of last remarks concludes the book. Those remarks are neither a summary nor a conclusion of the whole book. The way the chapters are structured does not require either for the readers. Final remarks are meant to stress and emphasize a few notes the author would like to leave readers with.

## Guiding notes for the readers

One thing that was considered in this book, but then dismissed, was unifying some terminologies that are widely used in the relevant literature. For example, in the development literature there are many names for developing countries: sometimes they are called the third world, sometimes low-income countries (or low and middle-income countries), sometimes less (or least) developed countries (LDCs), post-colonial societies, and sometimes the economic South.[16] The same for developed/industrialized societies: sometimes called the economic North, sometimes industrialized countries, developed countries, and other names.

This plurality of terminology applies to other phenomena in the field of development, itself a wide field that envelops aspects of many other fields of inquiry. While some may argue that each term within a category does not mean exactly the same thing as another, and while I can in principle agree with that, that level of specificity does not often apply to universal phenomena and arguments within the field. I opted for keeping this diversity of terminology in the text and using terms deemed suitable in different contexts. I did not see a major point in choosing and

---

16    The term "economic South" was coined by the Non-Aligned Movement countries
      to describe the bloc of countries who share the common experience of development
      challenges, a history of colonization and an economic dependency relation with the 'first
      world' countries (or economic North). The term encompasses countries that belong to
      this category but are located in the northern hemisphere, and vice versa, by distinguishing
      'Southerness' here as an economic-historical identity, not necessarily geographic, yet it is
      also the case that the majority of these countries are actually located within the southern
      hemisphere. For the record, I think this term is probably the most relevant for the themes
      of this book, but nonetheless I prefer to be able to use the other terms where I see suitable.

committing to limited terminology for the entire manuscript. On the contrary the allowable freedom to use a variety of terms was felt to be to my advantage and not deemed to negatively affect the flow of main points and arguments.

A big part of this work is an amalgamation of many writings of the recent past. Particularly, the technological autonomy conceptual framework and the core literature review about technological change and development draw on the author's doctoral thesis and research. Other parts of this book consult, and draw from, scholarly papers and opinion essays by the author, some published before, in various media, and some yet unpublished. All sources, however, were modified to various degrees to comprise a single manuscript format. That being the case, the book's language and arguments will reflect this. Clear and accessible language, however, were a main goal and I hope this was achieved sufficiently.

The book also tried to strike a balance between generalizations and details. It would not have been sensible to produce a detailed and jargon-filled treatise to share ideas on a topic that should be as public as possible. Similarly, it would not have made sense to generalise, without evidence and rigour, about a topic that requires both, and thrives on both, at its core. I hope I have succeeded in this aspiration.

Additionally, while the book addresses a problem relevant to all developing societies, readers will notice a tendency to address the multiple aspects of this problem through a lens and using examples mainly associated with, but not limited to, Africa. There is no oddity in this, for it is common that writers who address global problems – such as in the development literature –communicate on global platforms but draw mainly on experiences that are associated with regions of the world they are more familiar with, be they South Asia, Latin America, the Caribbean region, or the Middle East. The main issues are common; the details vary but are not totally alien to other contexts.

A last note: occasionally it is important to state the obvious. In that spirit I assert that this book could not have addressed all the aspects of technology, development and liberation, neither in breadth (i.e. the broadness of topics that belong to the spectrum of technological development) nor in depth (i.e. the layers in each topic). Yet the book sought to deliver a comprehensive argument for a keen approach to technology—a liberation technology.

Chapter One

# Technology, Institutions and Change

*"Our conceptual distinction is vital for any understanding of the interdependence of technology and institutions as well as their relative independence."*

–Karl Polanyi[1]

This chapter presents a literature review of the theory of technology, technology-institutional dynamics, and technological change. It explores, discusses and synthesizes the existing literature on technological change and development relevant to the framework of technological autonomy. It also provides a review of some of the major models of technological change along with their strengths, weaknesses and overlaps. The theoretic conclusions and syntheses reached through this chapter contribute to reinforcing the propositions and arguments of the introduction that laid the main thesis of this book.

Technologies only become functional in societies through institutional arrangements (formal and informal). They also tend to shape their encompassing institutions as they are shaped by them. Such dynamics manifest in processes of technological change in various societies according to various conditions that tend to compliment or run counter to each other depending on the contexts. In developing societies, where economies are less industrialized, technological change processes are mainly influenced by three conditions: technology-

---

(1)    In "The economy as instituted process", 1957, pp. 249.

institutional dynamics, the dichotomy between traditional and modern technology in key sectors, and development priorities.

## Defining technology

Defining technology is critical to understanding its role, components and expressions in contemporary societies. Incoherent definitions could obscure the issue, rather than illuminate it, or lead to incoherent conclusions. Coherent definitions of technology, on the other hand, can be multiple, as we shall see, but each has to be consistent as well as suitable for the context in which it is used.

Encyclopaedia Britannica defines technology as "the application of scientific knowledge to the practical aims of human life or, as it is sometimes phrased, to the change and manipulation of the human environment." The Greek origins of the word are the two words *techne* and *logos*, with the first meaning 'art, skill, craft' and the second meaning 'expression of'—rendering a somewhat literal meaning of 'expression of skill, art, and craft'.

Writings that focus on technology and development, and technological change, have provided various and keener definitions of technology. Attempts at definition that tend to be so broad as to include every aspect of the subject matter run the risk of being too broad to be called definitions.[2] If research on topics of technological change depends on such broad definitions of technology, it will likely have trouble avoiding vagueness. "To define, after all, is to set limits."[3]

Everett Rogers (2003, 13) proposes a conceptual definition: "A technology is a design for instrumental action that reduces the uncertainty in the cause-effect relationships involved in achieving a desired outcome." According to Rogers a technology would typically have two components: (1) a hardware aspect, represented by the material or physical embodiment of a tool; and (2) a software aspect, represented by the information base for the tool. An example for a technology that can be fluently described by this definition is the computer, where the hardware component is represented by the semiconductors, transistors, frame, etc., while the software component consists of the coded commands, instructions, and the package of operative information that allow utilization of the computer to achieve certain outcomes. Other technologies of various degrees and kinds can also be described by

---

(2)    See, for example, Haug 1992 and UNCTAD 1981.
(3)    Bhattacharyya 2004, 9.

Rogers' definition, such as animal-drawn ploughs—while the hardware is represented by animal power and the plough tool itself, the software aspect is represented by the human knowledge that combined and utilized the hardware in a particular arrangement to achieve a certain outcome. The articulation of "a designed instrument for uncertainty reduction in the cause-effect relationships" gives Rogers' definition a calculated conceptual elasticity, making it useful across the wide world of technologies but being precise enough to avoid the vagueness that renders some definitions useless.

Johan Galtung (1979) offers another robust conceptual definition, critiquing Rogers' approach to a degree. According to Galtung, technology generally consists of techniques and structures (as opposed to hardware and software in Rogers' definition). The techniques are the tools and know-how, while the structures are the social relations, or modes of production, within which the techniques are operational. Here we see that Galtung steps into the social dimension of technology that Rogers' definition did not necessarily include:

"A naive view of technology sees it merely as a question of tools (hardware) and skills and knowledge (software). These components are certainly important, but they are only the surface of technology, like the visible tip of the iceberg... Underlying knowledge there is a certain cognitive structure, a mental framework, a social cosmology, serving as the fertile soil in which the seeds of a certain type of knowledge may be planted and grow and generate new knowledge. And in order to use the tools, a certain behavioural structure is needed. Tools do not operate in a vacuum; they are man-made and man-used and require certain social arrangements to be operational." (Ibid, 6).

At a higher conceptual level, Galtung presents the axiom that technology constitutes the difference between natural ecological cycles and 'man-made' economic cycles. Thus, technology implies human modification of the natural ecological cycles. Therefore technology, in essence, is the modification of ecological cycles into economic cycles. While this definition is neat and concise it seems to be restricted to technology of economic utility only, unless we understand economic cycles in this definition, more or less, as virtually all ecological cycles that were humanly modified from their original/natural state to meet some human demand.[4]

---

(4)    While the general purpose of technology could be perceived as the betterment of human conditions – such as technologies of extraction, of transportation, communication, production, consumption, and of ecological balance – Galtung points out that not all

Ursula Franklin (1989) offers another definition that could offer an elaboration on other facets of technology: Technology is a system, a formalized practice, which includes ideas, practices, myths, and various models of reality, and it forcefully changes the social and individual relationships. Franklin then points out what technology is not—a helpful approach for 'setting limits' and making definitions clearer: "Technology is not the sum of the out-effects; of wheels and gears, of rails and electronic transmitters. [It] is a system [that] involves organization, procedure, symbols, new words, equations, and most of all it involves a mindset." (1989, lecture part 1). What is problematic, however, in Franklin's definition is that it seeks to exclude what she calls 'the out-effects'. After all wheels and gears, rails and electronic transmitters are the 'stuff' of the system she talks about. More precisely, the procedures, symbols, new words, and equations that comprise technology, in Franklin's definition, are so because they work on actualizing and operating the 'out-effects' that she downplays. The coherent point to be taken from Franklin, however, is that the out-effects could not alone define technology and must be seen as part of the whole package – the system – within which they are created and operated.

Borgmann (2010, 31) defines technology as "the transformation of reality according to the device paradigm, and it constitutes an ontology that is historic, dynamic, and enclitic or parasitic". What is referred to as 'the device paradigm' is generally the plethora of technological devices that pattern and have patterned our modern lives. Constituting an ontology of its own follows in this view from its being transformative of reality, where the tendency is "to mechanise and commodify the world at a certain place and has been spreading and sweeping everything before it since" (ibid). Borgmann's definition may provoke deep thought about the role of technology in shaping our contemporary reality but it fails to explain what a technology is composed of and what the functions it serves (as well as the context of its operation) are.

Moving further, Aunger (2010) makes the distinction that, whenever we talk about technology, artefacts have to be involved to some degree. Artefacts seem to be the common denominator among discussions of types of technologies (and even technological education). Therefore, Aunger concludes, technology is "about interaction with artefacts in particular contexts of engagement." This narrowed definition cuts out

modifications of ecological cycles seem to achieve that; indeed some seem to do quite the opposite such as military technology, for example.

attempts to classify even forms of social organization as technology; a wise thing to do for the sake of avoiding vagueness. Artefacts are particular material entities of known use and are the centre of interaction between an animal and the material environment in ways that this animal sees as a useful means to an end. Aunger intentionally does not exclude non-human animals as technology users and makers in this definition meaning artefacts range from bird nests to beaver dams to skyscrapers.

In synthesis, and for the purposes of this book, we will use the definition of technology as *artefacts built and used to reduce uncertainties related to particular problems within particular structures.* The remainder of the book will use the term 'technology' and its derivatives in accordance with this synthesized definition, unless otherwise indicated.

Besides the generic definition of technology, which defines the whole phenomenon, there are many subsets of the technology category with their own definitions that are not contradictory to the main definition but extensions to it. Before moving to the next section a few such subsets are defined below:

- Clean technology: defined as technology that delivers products or services that reduce negative environmental impacts.
- Appropriate technology: generally defined as technology envisioned with a specific development context in mind. It seeks to be sustainable for that context in particular—i.e., locally manageable in terms of know-how and material resources. "Compared to conventional technologies, appropriate technologies typically are less capital intensive; more labour intensive [nonetheless appropriate technologies are labour saving in comparison to traditional methods of production]; less dependent on scarce foreign exchange for imported goods; and easier to maintain, operate and repair" (Hyman 1987, 36).
- Biotechnology: as per the United Nations Convention on Biological Diversity (UNCBD), biotechnology is "any technological application that uses biological systems, living organisms or derivatives thereof, to make or modify products or processes for specific use."
- Information and communication technology: refers to the whole range of technologies that provide access to information through telecommunication devices.
- Indigenous technical knowledge (ITK): "refers to the knowledge, innovations and practices of indigenous and local communities around the world. Developed from experience gained over the centuries and adapted to the local culture and environment,

traditional knowledge... tends to be collectively owned." (UNCBD 2007).

## Technology and institutions

Many writers on technological change agree with Galtung (1979) in that there is a necessary interconnectedness between social institutions and technology. Social institutions refer to socioeconomic regulations, behavioral norms, incentives and expectations that constrain and shape human relations.[5] These social elements are 'institutions' when they persist through time. Hodgson (2007, 67) defines institutions as "durable systems of established and embedded social rules that structure social interactions", and North (1990, 3) defines them as "humanly devised constraints to human interaction". To Rogers (2003, 26) institutions "define a range of tolerable behavior and serve as a guide or standard for the behavior of members of a social system". The elasticity of the term institution allows it to be used to describe social entities (such as organizations) as well as social arrangements that do not have a material representation (such as laws, values and norms). Institutions stem from culture, belief systems and established common rules of conduct and 'doing business' (written or unwritten). They are structures that embed for instance culture and laws, and thus to speak of "institutions" is to evoke the structures that host, we can reasonably argue, cultural institutions, legal institutions, economic institutions, etc., all as variations of social institutions.

Understanding processes of technological change requires an understanding of the interdependence of technological knowledge, material resources and social context. The concept of 'technological embeddedness' captures this interconnectedness. It is drawn from the concept of embeddedness in the work of Karl Polanyi and others[6], which refers to how economic activities and processes are usually dependent upon social institutions. Technological embeddedness implies that, to be sustainable, technological choices need to be compatible with the socioeconomic and cultural structures within which they operate. Rogers (2003, 15) suggests that the adoption of innovations is dependent upon "the degree to which an innovation is perceived as being consistent with the existing values, past experiences, and needs of potential adopters. An idea that is incompatible with the values and norms of a social system

---

(5)  Menard et al. 2005; Orrnert 2006; Hodgson 2004; Voss 2004.
(6)  See Polanyi (1944, 1957 and 1968), Hopkins (1957), Dalton (1990), Harriss (2003), and Hyden (2006).

will not be adopted as rapidly as an innovation that is compatible." He refers to this condition as "compatibility". For example, Shaw (2002) demonstrates how educational structures and culture in the Arabian Gulf countries hindered proper introduction of advanced educational technologies to the region's schools.

The combined impact of technology and institutions on societies is difficult to miss, yet also difficult to articulate. To Polanyi, the economy itself "consists of technology employed within a context of institutions. This context is one of dynamic interaction. Institutions mould technology and technology moulds institutions." (Stanfield 1990, 203-4). Before Polanyi, Karl Marx put technology at the centre of his economic analysis of history, especially the history of the industrial revolution in Europe. Rosenberg (1982, 34) argues that the main reason behind "the fruitfulness of Marx's framework for the analysis of social change" was that "Marx was, himself, a careful student of technology." Indeed Marx was one of the early writers to acknowledge the social consequences of technological change. To a large extent, he saw the material forces of production as technological, while the relations of production were social:

> "Technology discloses man's mode of dealing with Nature, the process of production by which he sustains his life, and thereby also lays bare the mode of formation of his social relations, and of the mental conceptions that flow from them." (Marx 1887, 352)

From a general historical-materialist perspective, Marx (1977) suggests that changes in the material forces of production render changes in production relations. Marx's statement, in *The Poverty of Philosophy* (1847), that, "The handmill gives you society with the feudal lord: the steam-mill, society with the industrial capitalist" captures his general approach to the topic of technology-institutional dynamics. In a capitalist society, however, the aim of developing machinery, in Marx's analysis, is a socioeconomic aim—i.e. to increase surplus value (Marx 1887). This analysis seems to be consistent with Polanyi's (1957) conclusion that the main difference between capitalism and socialism is how modern technology is instituted in society. (However, Polanyi (1944) gravitated towards support of the argument that economic and political institutions tend to precede the environment that calls for certain technological shifts—i.e. social change precedes and determines technological change. Marx meanwhile seemed to attribute big social changes to technological changes that happen before them—such as the introduction of the factory, as a workplace, was a result of the introduction of the steam engine).[7]

There are strong correlations between the economies of land and labour and the process of advancing agricultural mechanization. Some writers emphasize the role of technological arrangements as the catalysts that necessitate and push for large social structures, such as states—i.e. the large geopolitical entities governed with strong centralized authorities. Karl Wittfogel (1957) presents a theory to explain how large governance systems (especially despotic ones of early times) were formed mainly to respond to problems of irrigation systems and water resource management. He called these states 'hydraulic states'. How the hydraulic state differs from the modern 'total managerial' state, Wittfogel (ibid, 48) argues, "is that it is based on agriculture and operates only part of the country's economy". Wittfogel relates this type of historical state to Asiatic societies and what has been called 'the Asiatic mode of production' to which ancient pre-colonial India and China – and ancient Egypt – belonged. If Wittfogel's historical analysis is credible enough then we can conclude from it that technology-institutional dynamics were strongly influential even in early times before the era of modern technology.

The main point emerging from the above arguments is the notion that there is a dynamic interdependence between technology and institutions and that technologies require a level of social embeddedness to be functional and continuous. Technological change requires institutional support, which means technological development is a social challenge. A 1993 report on technology transfer by the Organization of Economic Co-operation and Development (OECD) asserted that "the capacity for managing technological change is basically societal in nature, that is, it must permeate through many public and private institutions, all levels of society and be absorbed into the general "culture" of a country." (p. 4).

---

(7)     Worthy of mention here, as well, is Karl Polanyi's younger brother, who was a scientist and philosopher in his own right, and who also wrote on the topic of modern science and technology and society. Michael Polanyi is famous for conceptual contributions to understanding how scientific discoveries are related to cognitive and cultural drives for exploration through the concept of 'tacit knowledge'. The two Polanyi's had their agreements and disagreements on the details of understanding the unfolding and consequences of technology-institutional dynamics (see Kari Polanyi-Levitt's (ed.) *The life and work of Karl Polanyi*. 1990).

# Dichotomy of technologies: traditional vs. modern

In developing societies, development processes find themselves entangled in conflicts between two technological paradigms—old and new (or traditional and modern). The first is often the creation of technology users themselves, while the second is often the creation of designated professionals in modern societies, namely engineers and scientists, who are often generally separate from the technology users. For example, farmers do not design and build modern agricultural machinery, agricultural engineers do. This divide between creators and users does not exist in such a way with customary/traditional technologies. This divide between creators and users does not hold with traditional technologies.[8] Yet, for the sake of sustainable development, traditional and modern technologies should coexist and sometimes compete in a healthy manner. The divide between them thus poses a challenge for technological embeddedness, and may result in delays in the adoption of new technologies even if they are objectively more materially beneficial than the old ones.

Old local technologies and techniques are often described as 'traditional' or indigenous technical knowledge (ITK). In the historical cases of ITK, the lines are blurry between what constitutes technology and what constitutes cultural institutions (i.e. technological embeddedness is well established). This can explain part of the challenge of technological change in developing societies. For example, while agricultural mechanization in Africa needs to find a way to either replace or coexist with traditional agricultural technologies that are already embedded in local socioeconomic institutions, ICTs do not have to deal with that complexity. ICTs do not face similar resistance to adoption, compared to agricultural mechanization, because they are not replacing any technology that is already embedded in local socioeconomic institutions. The difficulty of replacing an existing technology that performs the same function could be a reason for the slow rate of adoption of new agricultural technologies. This line of argument generally asserts that existing social institutions could be significant barriers in technological change processes. The effects of cultural institutions on technology seem to only be highlighted when technology diffusion or transfer fails, for several reasons, among which may be the cultural unsuitability of the technology.[9] When cases of technological change are reasonably

---

(8)    Gamser 1988; Roy 2002; Visvanathan 2004.
(9)    See, for example, Eisler 2002; Rogers 2003; Adeel et al. 2008; Dengu et al. 2006).

successful (see, for example, Lekoko et al. 2012; Fidiel 2005; Gulrajani 2006; Gilbert 2009; Al-Ghafri 2008) there is little attention given to how accommodating the local values and beliefs played a role in making that technology adoption a success. When technology users who are adopting a new technology are in considerable agreement with the technological change, the cultural suitability in question is not given particular attention. What the literature suggests is that people's cultural institutions are relevant to the genuine participation of technology users in the technological change process.

While in modern societies there is a common social rhetoric of separating technology from culture, norms and rules, that rhetoric does not stand up to scrutiny. Galtung (1979), for instance, argues that there are two categories of social structures, alpha and beta structures, within which technology operates. The alpha structure refuses to recognize, and may not support, an embeddedness of technology in culture and nature, while the beta structure recognizes and assures such embeddedness. The alpha and beta structures tend to conflict when they share a social context, unless careful mitigation is applied to the process of technological change. This tendency toward conflict can explain part of the challenges of technological change in developing societies. Using the dynamics of technological embeddedness, mentioned earlier, it could be argued that ICTs do not face similar resistance in adoption rates, compared to agricultural mechanization, because they are not replacing any technological sector that is already embedded in African socioeconomic institutions. The process of replacing or absorbing existing technologies that perform the same functions in that context could be a big or the chief reason for the slow adoption rate of agricultural technologies. Research could further verify the argument that institutions – including cultural institutions – could be significant barriers in technological change processes. For example, renewable energy technologies also tend to have less resistance to adoption and diffusion activities than modern agricultural tools.

Howes and Chambers (1979) synthesize expert opinions that conclude the need for strong cooperation between ITK and modern science in the quest for development. Lucena et al. (2010) and Lekoko et al. (2012) also highlight the essential need for this type of cooperation. Scott (1998, 6) adds the concept of Métis to the literature of ITK, which generally means 'practical knowledge', and "denotes the knowledge that can come only from practical experience". "Far from being rigid and monolithic, [Métis] is plastic, local and divergent." (ibid, 332). The lack of access to scientific methods, laboratories and writing, did not mean that communities that relied on Métis lacked reliable knowledge, but rather only meant that they documented their rich and "remarkably accurate" knowledge systems through oral traditions and apprenticeship institutions passed from generation to generation. Scott criticizes development theories adopted by modern state-sanctioned schemes (so called top-down) that place large faith in modernization and modern products and services while dismissing the contribution of Métis from indigenous and native populations. Métis are also versed in its own techniques of experimentation and improvement (or their own R&D). A story from Nigeria demonstrates well this R&D capacity of ITK:

> "....a scientist believed he had made a breakthrough when he found a way of breeding yams from seed, propagation normally being vegetative. A farmer was casually encountered, however, who had not only himself succeeded in doing this, but had also discovered that whereas the first generation of tubers were abnormally small, the second and subsequent generations were of normal size. The scientist reportedly exclaimed 'Thank god these farmers don't write scientific papers'. It was also noted, in support of the prevalence of experimentation by farmers, that there is a Yoruba word for 'experiment'." (Howes and Chambers 1979).

However, Métis is not without its pitfalls, such as that (1) it is not democratically distributed, since it is passed through generations by means of direct apprenticeship that prefers kinship and exclusive leagues of artisans; and (2) it survives today more by isolation from the rest of the world than engaging it, which means its survival favours the isolation opted for by a few members of indigenous/native communities, not necessarily by collective communal choice. These pitfalls can be mitigated through dynamic interaction with scientific methods, but that interaction needs to be initiated first.

In further support of Scott's critique, studies such as that in Adeel et al. (2008) and the United Nations Environment Programme (2008) conclude that traditional water technology systems are not achieving their promised potential in contributing to technological development (and to climate change adaptation), due to two biases: 1) scarcity in scholarly research dedicated to evaluating and improving traditional systems, and 2) less favour by funding agencies (public or private) for R&D projects that aim towards improving the status of traditional ITKs (or Métis). Thus, for technological development schemes it is quite important to prioritize which sectors demand the introduction of new technologies and which sectors require more support and local upgrading for existing traditional technologies.

## Development priorities

When people get better at utilizing a technology they also often get better at realizing how they can improve it. Hence, learning and improving by practicing is an essential part of technological change. Rosenberg (1982) identifies a number of factors he calls 'supply side problems' that determine the rate of new technologies replacing old ones. The first factor is the continuity of inventive activity. Rosenberg observes that diffusion of technologies tends to be relatively slow everywhere with wide variations in the degrees of acceptance of different inventions. If we consider the steam engine a primary invention, for example, we can observe many 'secondary inventions' that improved the steam engine over a considerable period of time (Rosenberg 1972).

(Consider the story of the steam engine for a moment. It was patented by James Watt in 1781. It wasn't the first concept of a steam engine ever built; others were built decades before it. However it was an improved concept, and the one that began to show serious potential for the machine. Yet the machine did not begin to play a visible role in powering the British economy until the 1830s and 1840s; during that period many modifications to the original design took place. As late as the 1870s less than a million horsepower was generated by steam in the factories and workshops of Great Britain. Only by the 1870s and 80s was the steam engine ubiquitous in British industries (ibid). Overall, it can be said that the industrial breakthrough of the steam engine is actually a story that took almost a century to unfold.)

The second factor Rosenberg talks about is the development of technical skills among users. For example, it takes time to get labour to adopt or perfect a new machine. Only then will it be possible to demonstrate the extent of the quantitative and qualitative superiority of the new technology over the old one. The third factor is the journey from conceptual solutions to working machinery. With the institutional separation between modern science and R&D circles, on the one hand, and machinery making circles, on the other hand, it often happens that conceptual solutions take long periods of time to translate into working machinery. Other factors, like the economic feasibility of mass production and connection with willing investors (public or private), play a part as well. The fourth factor is the importance of the enabling environment for supporting innovations (i.e. macro policies and resources). An example is the role of the Japanese government in fostering Japan's automotive industry, which started with reverse engineering American automobile engines and saw the country grow to global leadership. Hence, successful technological development must be accompanied by a process of skills upgrades for both makers and users. This is one dimension of increasing technological capabilities.

Additionally, technology adoption stories may express themselves differently according to the technology sector and a society's priorities. Binswanger (1986) offers some insights on this issue through focusing on agricultural mechanization, and proposes generalizations about the multiple factors that influence transitions toward agricultural mechanization in different parts of the world (such as the US, Japan, Pakistan and Brazil). Three of the relevant highlights Binswanger presents apply to more than agricultural mechanization alone. The first highlight is on how the economy of land and labour endowments affects the priorities and pace of the process of shifting to agricultural mechanization. The second highlight is the relation between machinery design and capital costs: "Machinery design adjusts to high capital costs by lack of convenience features, simplicity, and reduced durability" (p. 36). While the cost of energy is a very important factor in machinery use, the costs of capital and maintenance tend to be usually larger. If capital cost is relatively high, then the innovator/adopter of the technology will try to maximize profit by compromising convenience, simplicity and durability. This highlight may explain why countries with least developed infrastructure tend to produce simpler, and less durable, engineered machinery—i.e. to compensate for the huge capital costs resulting from weak infrastructure to support industrialization (e.g., transportation infrastructure, consistent electricity and water coverage, industrial safety standards, strong vocational education) (Zanello et al. 2016). The third highlight relates to the decentralized nature of technology innovation. To the question of 'where does technological innovation often take place?' Binswanger has a direct answer: "In the early phase of [agricultural] machinery invention, subinvention[10] and adaptation are done almost exclusively by small manufacturers or

---

(10)    Perhaps what Binswanger refers to as 'subinvention' is the category of inventions that modify or change only parts of the one machine or unit rather than the whole machine. For example, the invention of the automatic transmission in automobiles, in the 1950s, was a subinvention in the sense that it did not change the automobile or the function of the transmission itself in it, but rather invented a new part that substituted an older one within the same machine. It is perhaps the same as how Rosenberg (1982) distinguishes between inventions and innovations. Other terms that could perhaps be used to describe

workshops, working closely with farmers. Public sector research has contributed little to machinery development, but more to education" (p. 50). Also, in agreement with Rosenberg (1972), Binswanger adds that "Inventive work on a particular operation often precedes by decades the widespread use of machinery. It reaches a peak during the initial adoption cycle, when derivative invention, refinements, and adaptation to different environments are required." (p. 51). Binswanger's observations on agricultural mechanization point to a country's context informing technological change patterns, and different patterns calling for different priority areas of technological development. Selected technologies may be chosen and given more focus for localization.

In summary, technological change experiences in developing societies are mainly influenced by three conditions: technology-institutional dynamics, the dichotomy of 'traditional versus modern', and development priorities in key sectors.

## Technological change models

We can broadly define the process of technological change as the socioeconomic and environmental transformation that takes place due to transformations in technology-institutional dynamics (discussed in the previous sections). The transformation can either be intentional – as a goal in itself – or an indirect byproduct of changes that occur in technology or institutions. This study is aligned with the approach to development that treats technological change as an intended goal, with both technology and institutions as means to that end.

The technological change literature proposes a number of models that explain the main features, movements and vehicles of technological change. Such models are not conclusive or universal but each model may work well in explaining certain cases. In other words, each model has its strengths and weaknesses. These models seek to illustrate general trajectories and mechanisms of technological change that integrate some of the multiple factors discussed above.

### Demand-pull model

The demand-pull model suggests that economic demands (not necessarily other forms of demand) and their transformation in response to different economic dynamics, are the main stimulus for

---

subinventions are: upgrades, derivative inventions (a term Binswanger also used in the same manuscript), secondary inventions, or even innovations.

new technological innovation and adoption (Dosi 1982). Studies that have been undertaken in support of this model show good evidence in some contexts. Ruttan (1997, 1520) narrates the story of a classic study, from the 1950s, of "the invention and diffusion and hybrid maize", which "demonstrated the role of demand in determining the timing and location of invention". Another study, from the same era, of patents statistics of innovations in multiple industries (railroads, agricultural machinery, paper and petroleum) "concluded that demand was more important in stimulating inventive activity than advances in the state of knowledge." (ibid, 2520).

In the demand-pull model – which itself has many versions – R&D of new technologies, and improvements to existing ones, are induced by economic demand pressures, which include competition, increasing demands, and external or environmental shifts that change demand in the market from one direction to another. For instance, Arthur (1989) proposes that competition between technologies in the market renders an 'increasing return': "Complex technologies often display increasing returns to adoption in that the more they are adopted, the more experience is gained with them, and the more they are improved" (p. 116). Competition in the market also highlights competitive advantages of different versions of the same technology, which will then promote the improvement of technology to respond to the demands of adopters. According to Arthur (1989) sometimes earlier access to some markets may determine the course of technological domination.

This model is not particular to developing societies, of course, but the implication of it for development studies is that to stimulate technological development it is necessary to stimulate local economic demand. This suggests that promoting technological change can largely happen through market incentives and regulations. This model says little about where the technology comes from (i.e. local or imported) and how sustainable that is. Oil-rich Middle-Eastern countries can be said to be technologically advanced from this perspective, while it is evident that there are yet big deficiencies in terms of national technology and science education and innovation in these countries (Pink 2009; Shaw 2002). In such cases the technologies have not really been localized, and the countries' own technological capabilities cannot be said to have improved proportionately over time. The demand-pull model also seems to take human creativity for granted. It provides no clues as to how to nourish and support innovative talents within developing countries to advance technologically. Furthermore, this model suggests that where economic demand stabilizes, or competition is non-existent, technological change will likely stall. This suggestion would be challenged by a body of historical economic studies of pre-modern capitalist, and older societies across the globe, where market forces did not seem to have contributed decisively to the many advancements – or regressions – that happened in technology and innovation (see for example, Polanyi et al. 1957). A number of modern technology breakthroughs were a result of the material and ideational sponsorship of the state's public sector without the initial triggering of demand, such as computer numerical control (CNC) machining (manufacturing automation) in the USA (Noble 1987). Rosenberg (1982) also gives an example of the technical change that happened in the commercial aircraft industry in the USA over the fifty-year period between 1925 and 1975. While it is by many measures a very successful story of rapid technological advancement, productivity and economic growth, the role of the federal government was the most important factor. Generally, governments of industrialized economies fund science and technology R&D generously through military and civilian research, and much of that funding renders technological breakthroughs that later 'trickle down' to the civilian markets (both national and global) (Mazzucato 2013), but these breakthroughs are not necessarily triggered by existing economic demands. Evidently the demand-pull model works quite well in many cases but not all.

## Technology-push model

In the technology push model, technological change is instigated by the innovative talents of individuals and teams. This model has a number of characteristics. First, there is the increasing connectedness between objective, non-market driven, scientific inquiry and technology innovation processes. Second, there is "the increased complexity of R&D activities which makes the innovative process a matter of long-run planning" (Dosi 1982, 151), for private and public organizations, which further distances the innovative process from direct market response. Third, while there is clear correlation between R&D efforts and "innovative output" (which can be measured by the patent activity, for example), there is not a similar clear correlation between market demand and the same measures of innovative output. Finally, as processes that seek to unfold what is not yet known, innovative processes are naturally surrounded by uncertainty (i.e. whether the R&D process, or scientific inquiry, is going to find solutions to the posed problems or not, or not-yet, etc.).

The technology-push model can explain how, for example, Singapore's massive investment in technology and science education helped transform it, in a few decades, from a poor country to a technologically advanced one (Patterson and Bozeman 1999). This model can also explain better the path of the Japanese Toyota company that made leaps in automotive engineering until it became a world leader in the manufacturing field with the Toyota Production System (Monden 1993). The development policy implications of this technological change model focus on innovative talents. It is important to invest in building and sustaining the proper institutions for nourishing technology and science and encouraging innovation. Yet, the shortcomings of this model seem to be the same advantages of the demand-pull model. At some point, innovative processes must be connected to the larger economic cycle to diffuse in society. The technology-push model says very little on this aspect.

## Path-dependence model

This model cites a number of historical case studies to emphasize how technological change is built up with every step dependent on the steps taken before it. The model explains historical developments. It also suggests that innovative talents and economic incentives are sometimes not able to save technologies from locked-in trajectories. Path-dependence can be illustrated by the case of the QWERTY keyboard (a descendent of the QWERTY typewriter). Rogers (2003; 8-10) gives a good historical account of how a more efficient typewriter (the Dvorak) was ignored by the vast majority of computer users due to a series of historic events that saw the QWERTY typewriter presented as the sole temporary solution to reduce the speed of typing (and thus reduce the number of mistakes printed). The evolution of the computer keyboard was locked into a system of path-dependence inherited from the slower keyboard layout. Today we do not need to worry about slowing down our typing speed for the sake of printing fewer mistakes, because we type on word processors and correct mistakes before printing to paper (and we may not even print at all). But still the majority of computer users around the world use the QWERTY keyboard and not the faster Dvorak. Ruttan (1997) also mentions the QWERTY case, and says that this particular case has attained the status of "a founding myth" in the path-dependence literature.

Rosenberg's observations (1972) also give credit to the path-dependence model. He argues that most technological innovations appear incrementally, diffuse incrementally, and depend on many external factors in the process.[11] He then presents a historical review of some of the outstanding technological changes that define our industrial times, such as Watt's steam engine (patented in 1781) and how its path to prominence took decades rather than years, with many incremental modifications and adoptions along the way.

Rosenberg also follows the incremental improvements of the steam engine, and how they needed other inventions to appear before they took place (such as the invention of more accurate-measure cylinders that helped James Watt decrease significant loss of steam from his engine). Technology improvement involves a learning curve that takes years, if not decades. During this time many events and incremental changes accumulate, without which the final product – if there is ever such a thing – does not stand. The early versions of any technological innovation tend to change dramatically over time, often with incremental changes contributed by various sources.

Multiple writers argue that the three models above are not sufficient – especially without integration – in explaining technological change in its generality (see Table 1). Ruttan (1997) argues that the three models represent elements of a general theory of technological change that has not yet been invented. However, the likelihood that such a theory could be found is small and detracts from the more useful effort to understand the diverse mechanisms or factors that contribute to technological change. Each of the models reviewed above describes some mechanisms and factors. It is thus sufficient to see them as part of a larger framework that encompasses and connects them and adds to them. The models above, for example, do not take into account cultural resistance to new technologies (i.e. embeddedness).

---

(11)   While the point Rosenberg is making – the incremental nature of technological improvement – is not necessarily the same as path-dependence, the two are evidently related.

# Table 1: Summary of explanatory models of technological change

| Model | Main Claim | Implication | Blind Spots |
|---|---|---|---|
| Demand-pull | The nature of market demands, and their transformation in response to different economic dynamics, are the main stimulator for new technological innovation and adoption. | Promoting technological change needs to happen indirectly though stimulating the market, by inducing market incentives and regulations. | - Obscures the role of innovative talents.<br>- Says little about the substance of science and technology R&D processes.<br>- Does not account for the inherent uncertainty in the R&D inquiry process. |
| Technology-push | Technological change is autonomous, or quasi-autonomous, of market mechanism, has its own dynamics, and is highly instigated by innovative talents. | Promoting investment in building and sustaining the right institutions for nourishing science and technology and encouraging innovation. | - Fails to account for the importance of economic conditions (institutions, resources, demand, etc.) to complete the technological change cycle. |
| Path-dependence | Technological change is an incremental process that is path-dependent. New innovations build on the previous prevalent products and techniques (which are not necessary the best ones, but happened to be prevalent due to other social, economic, or peculiar circumstances). | Policy implications of this model are unclear. It helps explain the evolution of many forms of products and techniques, but does not quite indicate which development policy to apply. | - May cover a good range of technologies, but not all of them.<br>- Policy implications are less lucid. |

There are some other models that seek to draw a more comprehensive map, however. One example is Dosi's (1982) concepts of 'technological paradigms' and 'technology trajectories' as a more comprehensive approach than the models mentioned above. Technological paradigms are uniform patterns of solutions offered related to sets of technological problems, "based on selected principles derived from natural sciences and on selected material technologies." (ibid, 152). A typical technological paradigm would include a basic set of prescriptions to follow when addressing a certain 'cluster of technologies' (e.g. agricultural technology, transportation technology, energy technology, etc.). Technology trajectories are the 'normal', cumulative progress within the same technological paradigm (a continuity) while a shift to a substantially different way of innovation represents a 'discontinuity' of the paradigm, hence a new technological paradigm. People following a certain technological paradigm can excel in finding and implementing similar solutions to familiar problems, but they also tend to exclude other imagined possibilities, as solutions, in favour of normal – or normalized – expectations within the paradigm. Such exclusions might be due to limitations in the innovative thinking itself, the institutional environment, the scientific abilities, or a combination of these. Once in a while come dramatic events that were either instigated by innovative talents or by institutional transformations (intentional or accidental) that cause significant changes in technological paradigms; allowing new paradigms to emerge.

Sometimes, however, a technology trajectory can also change – from weak to strong and vice versa, or from slow to fast and vice versa – if significant changes happen in the same cluster of technologies, but without causing a change in the main, technical and institutional, principles of the technological paradigm. An example for this change in trajectory, but not in paradigm, is the shift that that happened around the 1950s with the introduction of automatic transmissions to automobiles. With the introduction of the automatic transmission driving became more accessible to a wider range of consumers due to significant increase in convenience of operating vehicles.

Dosi's model of technological paradigms may prove useful in explaining some historical events, but does not seem capable of explaining contemporary processes of technological change in developing societies. For example, the technological divide that may exist inside the same developing country – between modern and traditional technologies, or urban and rural communities – is among the development challenges that are difficult for this framework to explain. In a developing country, such contradictions may look as if they are somehow all embedded, as far as this framework is concerned, but the status of technological dependency and its consequences on human development, are not flagged. Generally, Dosi's model explains changes in individual technologies but not the general movement of countries along the path of increased technological capacity or technological autonomy.

In summary, we can conclude that to describe what an effective technological change process would be for developing societies, the existing models of technological change are not on their own sufficiently broad. They offer points of guidance but need to be put together in a framework that provides a more coherent understanding of what the objectives should be and what are the challenges of national or sectoral technological development. In particular, a framework is required that identifies how technologies are localized by a developing society and how technological autonomy is pursued and attained.

## Conclusion: understanding technological change

As a collection of artefacts that are built and used to reduce uncertainty related to particular problems, technologies only become functional in societies through institutional arrangements (formal and informal). The problems that such artefacts are utilized to resolve often emerge in the contexts of particular societies and environments. The various and complex technology-institutional dynamics manifest in processes of technological change, tend to support general development favourables (e.g. human development, economic development, industrialization, sustainability, etc.) when steered in favourable directions. In the contexts of developing societies, where economies are less industrialized and technology dependency reflects in unfavourable human development indicators, we recognize that technological change processes are mainly influenced by three conditions. These conditions are technology-intuitional dynamics, the development priorities of the society in question, and the dichotomy of traditional vs. modern in key sectors of technology.

There should be a sort of a diagnostic approach to technological change in developing societies. A diagnostic approach implies that although each social context is unique it also shares common attributes and criteria of analysis with others under the same category; analogous to diagnosis in the medical field where human beings need to be treated as unique cases but with the understanding that they all share common organs and functions.[12]

Figure 1 shows a possible visualization of the definition of technological change. It is just one way of visualizing what constitutes the basic components of technological change (i.e. not a description of the process itself or the path towards autonomy followed during that process, which is the topic of the next chapter).

**Figure 1: Visualization of the definition of technological change**

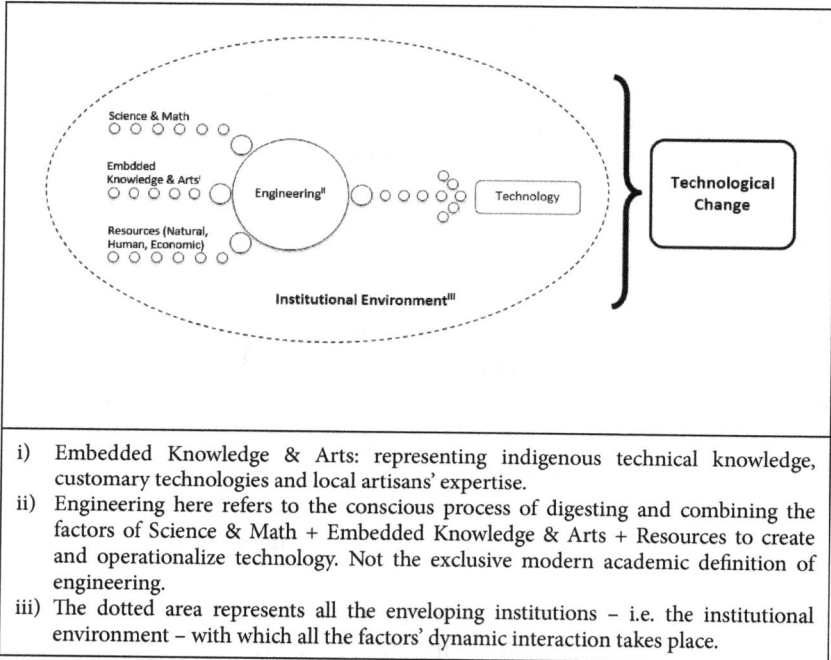

i) Embedded Knowledge & Arts: representing indigenous technical knowledge, customary technologies and local artisans' expertise.

ii) Engineering here refers to the conscious process of digesting and combining the factors of Science & Math + Embedded Knowledge & Arts + Resources to create and operationalize technology. Not the exclusive modern academic definition of engineering.

iii) The dotted area represents all the enveloping institutions – i.e. the institutional environment – with which all the factors' dynamic interaction takes place.

---

(12)    Basurto & Ostrum 2009.

# Chapter Two

# Towards Technological Autonomy

*"You never change things by fighting the existing reality. To change something, build a new model that makes the existing model obsolete."*

–R. Buckminster Fuller

Technological autonomy refers to a society's attainment of a sufficient level of endogenous capacity for generating, transferring and adapting technologies, guiding industrialization and innovation, and setting technological development priorities in order to achieve self-determination in planning and managing its technological affairs. The concept of technological autonomy has not been widely discussed in the technological change or development literature, but Morehouse (1979, 387)) has referred to it as a goal for developing countries:

> "Development strategies, relying on importation of capital-intensive, socially inappropriate, environmentally destructive Western technologies... have been at the heart of the accelerating de-industrialization of the Third World... While we cannot be certain that greater autonomy will lead to greater equity, few Southern countries can go very far in meeting the minimum material needs of most, not to speak all, of their people without a greatly strengthened autonomous capacity for creating, acquiring, adapting and using technology to solve their own urgent economic and social problems."

Morehouse (1979, 397) adds that, "technological autonomy is not, of course, autarky, but it does imply greater selectivity in, and closer control of, externally acquired technology". Technological autonomy is reached when self-determination of technological affairs is attained. This requires both increased technological capabilities and increased technology localization, which are concepts we elaborate about in this chapter.

## Technological dependency and autonomy

To develop the concept of technological autonomy we can begin by looking at its opposite: the state of technological dependency. During the 1960s and 70s many 'third world countries' sought to negotiate terms of technology transfer with 'first world countries' and transnational corporations, through the United Nations Conference on Trade and Development (UNCTAD), the New International Economic Order (NIEO),[1] and other multilateral organizations and frameworks. That experience was summarised by Haug (1992, 218):

> "The third world's frantic attitude toward technology transfer resulted in the countries falling victim to a sort of "technological colonialism." Taking advantage of the third world's desperation, TNCs [Transnational Corporations] drew-up largely one-sided transfer agreements. For example these agreements often linked the transfer of technology to the right to build, operate, and maintain the manufacturing plants. Suffering from a lack of information about the technology and about the transfer process, many third world nations accepted these agreements. Consequently little technology was actually transferred to the developing countries, and those countries failed to develop an indigenous technological capacity."

Haug lists a number of problems that made developing countries vulnerable to such 'technological colonialism': lack of reliable infrastructure conducive to optimal technology utilization, failure to develop local technological skills, importation of inappropriate technologies for local contexts due to insufficient information and knowledge, and absence of technological development plans (i.e. institutional and policy immaturity).

Desai et al. (2002, 97) argue that "Not all countries need to be on the cutting edge of global technological advance, but every country needs the capacity to understand and adapt global technologies for local needs." Anything below that capacity can be described as a situation

---

(1)   The NIEO was a program of action that was approved in 1974 by UN General Assembly, which was "intended to eliminate the economic dependence of developing countries, promote their accelerated development based on the principle of self-reliance, and introduce appropriate institutional changes for the global management of world resources." (Haug 1992, 219).

of technological dependency. They argue that many developing countries today are in that situation. Lall (1992, 182-83) has said of this phenomenon:

"Technological development always needs technology imports from advanced countries. The extent of dependence on imported technology and the form that technology imports take, however, affect NTC [national technological capabilities] development. A passive reliance on foreign skills, knowledge and technology may lead to NTC stagnation at a low level, while selective inputs of foreign technology into an active domestic process of technology development can lead to dynamic NTC growth."

Therefore, technological autonomy does not mean that a society is self-sufficient in technological products and services, without having to engage with the rest of the world. In this age of globalization, self-sufficiency is extremely difficult and, more importantly, gives no particular advantage. The technology supply chain and global market give no advantage to self-sufficiency, but to technological autonomy. "Autonomy" implies the ability to engage the rest the world in the exchange of technological products and services with a level of agency that does not make the society a helpless receiver of technology, without power to choose, negotiate, and have a degree of technological sovereignty.

## Proposed technological autonomy framework

Technological autonomy refers to the attainment of a sufficient level of self-determination over technological affairs for a given society. A "sufficient level" implies that there is endogenous capacity for making and executing decisions related to guiding innovation, industries, technology transfer, and priorities for development. For example, a country is technologically autonomous when its food security and basic infrastructure (housing, utilities, transportation, and basic telecommunication networks) are not threatened by supplier countries. Meanwhile it should also have autonomy in delivering the basic services of education and healthcare, meaning it has the capacity to build and reproduce technological knowledge and skills of the local population, and to foster their innovative talents, in an environment that also meets basic health and safety requirements, without being particularly dependent on external powers as suppliers of those basic services.

The framework, provided here, identifies a set of concepts and relationships that can be used to build theories and explanations about technological autonomy. It follows Ostrom's (2005) criteria for a conceptual framework, as that which is comprehensive enough to address the general problem – technological autonomy – and flexible enough to allow for the creation of multiple models and policy strategies, depending on the context, to achieve a broad goal. The technological autonomy framework can support the development of explanations for technological change processes, guide research on technological change, and contribute to policy formation, strategic planning and implementation. Additionally, the specification of the main variables of technology localization (diffusion, institutional support and adaptation) is useful in identifying research gaps and in mapping and selecting policies that suit various organizational, social, and jurisdictional contexts.

By highlighting the elements in technological change processes, the framework is intended to guide areas of focus for research clarity and informed decision-making. Both then contribute to conscious, strategic planning and implementation (i.e. praxis). The main features of the framework are shown in figure 2. It shows a technological change process leading to the achievement of a basic level of technological autonomy—i.e. to abate technological dependency. The main variables of that process are the internal factors of (a) technology localization, and (b) technological capabilities. While technological capabilities relate to the general enabling environment for a productive society, technology localization addresses chosen technologies of particular importance and priority in specific local contexts.

Different developing societies manifest different levels of technological dependency, as can be determined from their state of technological affairs. Moreover, their various contexts will determine various policy and structural planning and implementation priorities, based on their relative sets of assets and vulnerabilities. Therefore, different levels of technology localization and technological capabilities are expected (figure 2). A developing society (country, region, community, etc.) can be located at any point on the spectrum between technological dependency and technological autonomy, at any particular point in time, and can be moving forward (or backwards) on that spectrum. Technological development implies moving forward towards technological autonomy.

## Figure 2: Framework for technological autonomy

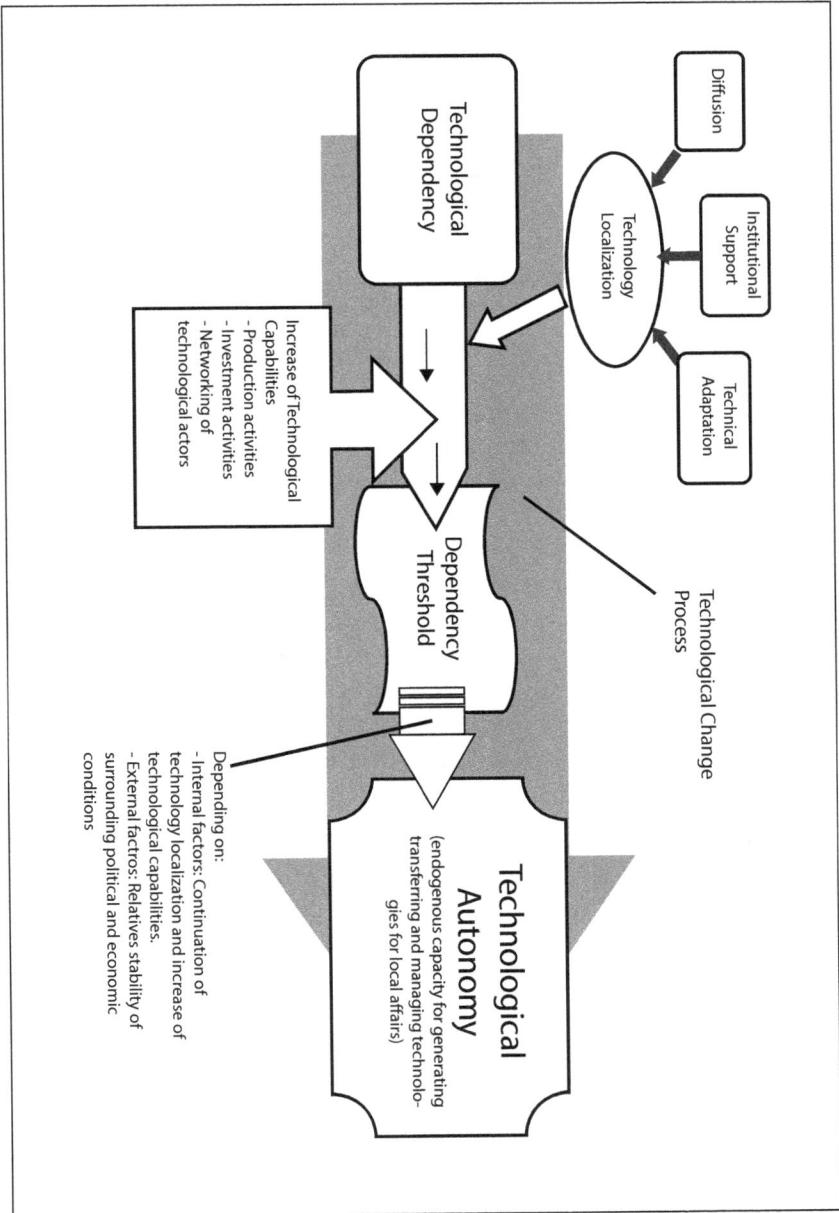

The dependency threshold in the framework (figure 2) suggests that if a society is beginning to free itself from the limitations of technological dependency and move towards technological autonomy, then we can say that that society is at a dependency threshold. This threshold state is made possible by internal and external factors that, together, can lead a society towards technological autonomy. The internal factors are represented by the activities of technology localization and the increase of technological capabilities (production activities, investment activities, and networks of technological actors). The external factors are represented by an assumption of relative political and economic stability. External forces have an influence on any process of technological change, even if the internal factors are optimal. However, the scope of the external factors is best treated separately from the focus of the framework, since it is a broader topic with more diverse variables.[2] Reference to external factors in the framework is important however so as not to suggest that only internal factors play a role in the attainment of technological autonomy.

## Increase of technological capabilities

Technological capabilities refer to the "dynamic resources which encompass the skills, knowledge and routines involved in generating and managing technological change, whether they concern production activities, investment activities, or relations with other [technological actors]" (Albu 1997, quoted in Gulrajani 2006, 154). Thus, the activities for increasing technological capabilities fall into three categories: production activities, investment activities, and the networking of technological actors (Figure 2). Production activities include the integration of skilled labour and natural resources in making and utilizing technologies. Investment activities concern decisions and actions in channeling finite resources into enhancing technological sectors by increasing human resources, accessing natural resources, or R&D endeavors. Networking of technological actors refers to the interactions and communication between producers, investors and innovators that coordinate their work and pursuits. Networks are known to be generally more productive and efficient than isolated or fragmented groups of actors. In other words, technological capabilities concern the general enabling environment for technologically productive forces in a society. The technological change literature demonstrates that effective technological change experiences,

---

(2)    This book addresses multiple external factors that influence technological change
        processes in subsequent chapters.

by and large, have been the outcome of multiple actors and institutions building upon each other's work (see for example Rosenberg 1972; 1982; Binswanger 1986; OECD 1993; Nasir et al. 2011).

The concept of technological capabilities was introduced by Lall (1992) and Gulrajani (2006) in the context of industrial clusters but was later enhanced and applied to firms as well as to national systems. Lall (1992) takes technological capabilities from the firm and industrial cluster level to the national level. Lall suggests a 'simple framework' for national technological change planning that includes: policies of market incentives (micro and macro), increasing technological capabilities, and institutions. By incentives Lall refers to the policy and market stimulants that encourage investment in innovative endeavors, on-the-job training and business environment improvement. By institutions Lall refers to the national bodies that plan and execute interventions, and regulate constraints, to induce economic, social and innovative factors to work together in investing in human and physical capital.[3] A similar conclusion is reached by others who examined the stories of industrial development in newly industrializing economies (NIEs) (e.g. South Korea, Taiwan, Brazil, South Africa, India, and Malaysia) (See Kim and Nelson's (2000)).

These stories of NIEs can be viewed as stories of transformation from technological dependency to technological autonomy. Yet the framework that Lall suggests does not address issues of diffusion and technical adaptation, particularly at the level of users and smaller social units within the national umbrella (such as rural communities). Additionally, Lall's framework does not sufficiently address two of the conditions that technological development faces in developing societies: the dichotomy of modern and traditional technologies and the context-determined priorities of development (which can seriously influence the trajectory of technological change in a given society). Such issues have a record of causing resistance to technological change among targeted communities in developing societies. The framework of technological autonomy addresses these issues through the variable of technology localization.

---

(3)    Note: the definition of institutions by Lall (1992) is not the same as the definition adopted by this paper. Lall's definition is particularly limited to national formal institutions.

# Technology localization

While the increase of technological capabilities represents the strategic, continuous and cumulative side of a technological change process, technology localization represents the interventionist side of the process. It is meant to respond to incidents where there are persistent elements of resistance to change regarding particular technology types. For a variety of reasons, some technology types suffer persistent low levels of adoption over time in developing societies (e.g., agricultural technologies in sub-Saharan Africa in the last 3 decades) while others do not (e.g., ICTs in sub-Saharan Africa during the last decade). In cases of resistance focused interventions may be required. Technology localization refers to activities that seek to make chosen technologies locally functional and locally embedded (i.e. without high resistance). Such technologies are typically of particular importance and priority in specific contexts. The main activities of technology localization are:

1. Diffusion; which refers to persuading the concerned segments of society, through communication and promotion, to adopt new technological innovations.

2. Institutional support; which includes policy advocacy, resource mobilization (e.g. finance and credit systems), and logistical and training assistance for the operationalization of chosen technologies.

3. Technical adaptation; which includes additional, incremental, technical modifications to the technologies of concern, or complementing technologies in ways that add value and increase usability of the technologies in specific contexts.

In brief, local technological change requires diffusion of new technologies, support for the technologies' operationalization, and sometimes adaptations of available technologies to fit more effectively into a local context. The three processes are mutually supporting. Technology localization, if successful, increases technological change and increases technological capabilities. To explain the contribution of technology localization to technological autonomy, we first have to look at the nature of the life cycle of technological innovations. Understanding that life cycle – from inception to maturity to diffusion and adoption – allows us to recognize the value of the role played by diffusion, institutional support and adaptation in promoting technological change.

Innovations emerge out of conditions where different objective and subjective elements come together in particular historical contexts.

Those contexts present incentives and constraints that are sometimes institutional (formal or informal) and sometimes environmental (i.e. ecological, geographical, etc.) (Rogers 2003; Hodgson 2004). Most innovations in history occur in decentralized settings where smaller groups or individuals share the credit for the innovation (Binswanger 1986; Rosenberg 1972). Even when macro planning is involved, it manifests in giving latitude, mission and resources to particular small groups, such as engineers, scientists, technologists and artisans (and their historical equivalents), to undertake innovative projects (Noble 1984). Innovation tends to be carried out in smaller entities in society, such as firms, laboratories, and specialized teams/associations (such as artisans and technicians). Even within those smaller entities, it is common to find even smaller teams (or individuals) as the initiators and custodians of new technological products and systems.

It is critical to distinguish between innovation, on the one hand, and innovation diffusion on the other hand. Innovation involves learning from conditions and environments, conceptualizing solutions, and providing material demonstrations of those solutions. Diffusion of innovation, which comes after the realization of the innovation, is "the process by which an innovation is communicated through certain channels over time among the members of a social system... The diffusion process typically involves both mass media and interpersonal communication channels." (Rogers et al. 2009, 418). Rosenberg (1972;1982) and Binswanger (1986) document that, generally, new technologies require a considerable amount of time between their first incubation and the time when they show a visible level of adoption. Noble (1984) and Kreszner (1987) document the time it took private manufacturing firms in the USA to adopt computer numerical control (CNC) machining from the time the technology was ready for adoption. Technology localization refers to the activities that take place in the period between the discovery of a technological innovation and the time it becomes widely adopted in society. The activities of technology localization can occur at macro as well as micro levels (see table 2). Diffusion is thus one element in localization.

The other activities of localization are institutional support and technical adaptation. Institutional support addresses the need some technologies present to have support in the form of advocacy, resources, training and logistics as they gradually permeate society and become integrated. Technical adaptation addresses the required alterations to technologies adopted from other societies to make them more compatible, in functional and sustainable terms, with the new local context.

Table 2 lays out a variety of sample technology localization activities at national levels, organizational levels and between. Each level has corresponding sample activities of technology diffusion, institutional support and technical adaptation.

While innovation has a decentralized nature, localization is a collective process. It needs more collective work because it addresses a challenge of a collective nature, that is to make a technology adopted and effective in a particular society. Developing societies face critical challenges in achieving technology localization.

**Table 2: Technology localization matrix (sample activities)**

| | Diffusion | Institutional Support | Technical adaptation |
|---|---|---|---|
| **National level** | - Promotion of hands-on training and applied research in educational institutes, etc.<br>- Establishment of national innovation systems. | - Infrastructure building (e.g. transportation, power and water, etc.)<br>- Policy incentives and constraints<br>- Local innovation-friendly taxation systems | - Provision of counseling and access to R&D information resources to organizations and communal initiatives<br>- Establish national R&D institutions for critical development sectors |
| **Communal/social level (municipalities, villages, civil society)** | - Use of mass media to communicate new technological solutions and highlight challenges to encourage ideas<br>- Promotion of innovative solutions through media local funding (municipalities, etc.) | - Schools to have embedded curricula of technological orientation (i.e. innovative problem-solving projects drawn from the surrounding environment, etc.)<br>- Worker associations to hold technological education seasonal workshops, etc. | - Community initiatives such as local competitions and awards for 'best innovations' for certain local challenges<br>- Technology incubators and innovation centers |

| | Diffusion | Institutional Support | Technical adaptation |
|---|---|---|---|
| Organizational level (firms, social enterprises, NGOs, etc.) | - Adopting organizational systems that are innovation-friendly and employee-empowering (such the Toyota Production System) | - Resource allocation for R&D teams<br>- Positive risk and opportunity assessment, etc. | - Apprenticeship programs, skills-upgrade courses, reverse engineering projects, etc. |

## Some elaborations

In economic terms, technological autonomy would also mean greater involvement of localized technology in value addition and creating the needed surplus to maintain local economic growth and expansion of public welfare. Without the creation of a sustainable surplus a local economy will likely stagnate and fail to transform the socioeconomic conditions of the masses.[4]

The relevant development studies and economic history literature shows that a multitude of contemporary societies with various contexts strive to engage a technological change process that sets them on the path towards attaining technological autonomy. Whether expressed explicitly or implicitly, each developing society can be shown to be on a quest to identify the proper variables and agents for its context to effect a successful technological change process. For example, a country can be technologically autonomous when its food production (agricultural and agro-processing) is no longer dependent on power relations with other supplier countries – resonant of the concept of food sovereignty proposed by *La Via Campesina* (International Peasant Movement) – or if its basic infrastructure (e.g. housing, utilities, transportation, basic telecommunication networks) is dependent on similar power relations. With the basic services of education and healthcare, technological autonomy would mean the capacity to build and reproduce technological knowledge and skills of the local population, and foster their innovative talents, in an environment that also meets the basic health and safety

---

(4)    We cannot dismiss the argument that the main indicators of a country's economic development (inflation, economic growth, employment, poverty, and equality) should always be accounted for. No matter what the political jargon is, if a country is failing in these indicators then it is most likely failing in its development aspirations.

requirements, without being particularly dependent on external powers as suppliers of those basic services.

Different developing societies manifest different levels of technology dependency, as could be determined from their state of technological affairs. Moreover, their varying contexts will determine policy and structural planning and implementation priorities, based on their relative sets of assets and vulnerabilities. Therefore, different contents and mixtures of technological capabilities and technology localizations are expected (figure 2). All in all, a developing society (country, region, etc.) can be located at any point on the path towards technological autonomy for a particular point in time, and can be moving forward (or backwards) on that path. The aim however is to be moving forward at a satisfying pace.

For example, it took Singapore about three decades to transform from a low-income country to a high-income one, with concurrent transformations in its economic cycles, infrastructure, education and health care. Patterson and Bozeman (1999) argue that it was particularly Singapore's investment in technology and science education and promotion that begat this transformation. The difference this made is put in stark relief if we compare it to other countries it was on the same development level as three decades ago – such as Botswana – but that are now a great distance behind it. China and Brazil are also good examples having emerged, through decades of work, from technologically dependent to technologically autonomous, industrialized countries. Other countries, such as India and Malaysia, are quite interesting cases to look at in terms of visible historical developments – with some salient challenges – along the path towards technological autonomy. There are even peculiar cases, like that of Cuba, which despite decades of imposed economic and political pressure from its neighbouring giant, the USA, persisted in maintaining high (and impressive) indicators of achievement in the healthcare sector, environmental sustainability and quality of education.[5] There are also cases to watch for closely these days, as they seem to be making interesting breakthroughs along that path, such as Vietnam, and a few African countries.[6]

---

(5)    Although there are legitimate concerns about the heavily centralized and authoritarian model of governance, that Cuba maintained such high development indicators is clear evidence that Cuba must have been doing some things correctly. Those things are worthy of learning.

(6)    There are surely many questions marks, at this point, about the consequences of the trajectories that countries like Ethiopia and Vietnam are taking (and they are by no means similar trajectories) that it will only be possible to answer in the future as we see

But in deeper analyses we can, and should, examine various sectors and processes within a social unit (i.e. country, region, community, etc.) that are relevant to technological autonomy variables and how they are doing.

The potential benefits of the proposed technological autonomy framework may not be seminal at the outset. Relatively new concepts for the analysis of technological change and development may have been presented thus far, but those concepts are essentially a re-arrangement and synthesis of ideas and values that have been expressed before in the same literature. What this framework would be concerned with, however, is adding value to the way the main variables of technological change processes can be integrated together to guide policy agenda (as well as critical study agenda). By developing a common language – common concepts and markers – to be used among those involved in technological change processes, and clarifying the relations between phenomena, the technological autonomy framework may help to collectively clarify and visualize goals and objectives of technological change processes that seek technological autonomy.

developments unfold.

## Chapter Three

# Agents of Technological Change

*"Power is the ability to define phenomena, and make them act in a desired manner."*

–Huey P. Newton

The framework of technological autonomy, described in the previous chapter, is led, operationalized and set in motion by agents of technological change. These agents are not 'in' the framework because they are not 'variables' in it; rather they are the ones that operationalize the variables. At communal, social, national and regional levels, these agents are seldom individuals (although sometimes they may be). They represent groups, associations and bodies, even when they act sometimes and in some respects as individuals. Different contexts require different agents of change, hence there are a variety of them, established or emerging, at various levels of engagement, aiming to stimulate and actualize technological change through such formations as national innovation institutions, educational institutes, technical training and R&D establishments, manufacturing and design firms, technology incubators, NGOs, and social enterprises.

In the technological change literature, much study has been made of the role of the state (public/national sector) and agents of the market: private firms and technical consultancies. There is also a range of combination of the two (public and private sector), such as public-private partnership programs, parastatal corporations, and transnational corporations with some public-sector shares. In varying degrees, the state and the private sector are active in industrialized, semi-industrialized and low-income countries.

Overall, it cannot be denied that states and markets host the most prominent agents of technological change in our times. That we shall speak about both in this chapter. It is critical to understand how, and under which conditions, states and markets fulfill their role in promoting technological autonomy.

## The state and technological change

The role of the state in technological change deserves special attention, due to the critical role of state institutions in our contemporary realities everywhere. In developing societies there is no evidence that any genuine measures of technological development have been achieved under weak or non-developmental states.[1] For example, Kim and Nelson (2000) show that in the newly-industrializing economies (NIEs) developmental states mobilized national industrialization plans utilizing the state apparatus and resources as a key agent of technological change. The state can furnish the enabling environment for industrialization, provide resources, and support the growth of key or immature industries. It can also support human capital formation and innovative technological R&D thus encouraging continuing growth in general or in particular sectors.[2]

The USA military and university engineering researchers played a paramount role, using public funds, in the development, introduction and persistent push to shift production processes towards adopting industrial automation (Noble 1984). Computerized numerical control (CNC) machines were not developed by private manufacturing companies, but rather were later adopted by them after considerable investment (financial and promotional) by the US military and research engineers (notably from the Massachusetts Institute of Technology). It is said that the intervention of the military in this big technological transformation was for political reasons (Kroszner 1987). The public budget that funded military and university engineering research for CNC machining was critically larger and minimally constrained than what any private sector company, of any size, would have been able to afford for an R&D project. Furthermore, there was no evidence that manufacturing industries expressed demand for CNC at the time.

---

(1)     See Nyerere 1998, Kim and Nelson 2000 and Nasir et al. 2011.
(2)     See for example, Noble 1984; Kroszner 1987; Wolff 1999; Ash et al. 2006; Mazzucato 2013.

Furthermore, the adoption of CNC machines by private industries was much slower than publicly perceived, because CNC machines required large capital investment that did not make economic sense at the time to private producers and investors. "According to *American Machinist*, less than 4% of all metal working machines in the United States were numerically controlled by the start of 1983."[3] Overall, numerical control "was brought into being not by the market, but by the state, at public expense."

Another example is the robust technological progress in the commercial aircraft industry in the USA over the fifty-year period between 1925 and 1975. While it is by many measures a very successful story of rapid technological advancement, productivity and market growth, the role of the federal government was the predominant factor. "For a variety of reasons, including the strategic military importance of aircraft and a concern with passenger safety, the federal government's role has been particularly prominent with respect to aircraft" (Rosenberg 1982, x). The heavy public regulation of the aircraft industry rendered good results for the commercial aircraft market. This may also draw attention to the continuing reality of how the USA federal authorities, and other governments of industrialized economies, fund science and technology R&D generously through military research, and how much of that funding renders technological breakthroughs that later 'trickle down' to the civilian markets (both national and global). Think the National Aeronautics and Space Administration (NASA)[4], for another example. The story of the development of the internet overseen by the USA department of defence is another prime and popular example. The role of the Japanese government in fostering Japan's automotive industry – starting early from reverse engineering American automobile engines – which grew into a prominent global leader today, is also worth mentioning.

Mazzucato (2013) also argues that the state has consistently played the leading and steering role in most technological advancements in modern societies, including in capitalist countries with a strong preference for a flourishing private sector. She demonstrates her argument through multiple examples, including how direct US state sponsoring and funding for innovative, risk-taking and breakthrough R&D projects in telecommunications and biomedical sciences played the major role

---

(3)     Kroszner 1987, 12.
(4)     NASA is an independent science and technology R&D agency of the executive branch of the United States federal government.

in giving birth to the internet, GPS technology and breakthrough discoveries in pharmaceutics. And for one extreme, but very legitimate, example, it was through the efforts of the state that the USSR took its society through a mammoth technological change process that, in only a few decades, transformed it from a largely agrarian economy to launching the first artificial earth satellite and sending the first human into outer space.

Scott (1998) highlights the other side of the coin of modern state intervention in technological change: the high costs of human suffering that mega projects by many states, caused, without achieving their prophesized goals to satisfying degrees. Examples of such stories are India's large dams, forced villagization in Tanzania, collectivization in the USSR, and the Great Leap Forward in China. Scott attributes certain qualities to state structure that make it inherently incapable of addressing problems on detailed and specific levels. The state 'sees' only large patterns and large actors and entities, and so it responds with generalized solutions. Scott says that he is not an enemy of the modern state structure, per se, but wants us to recognize its clear limitations, in order to give space to more localized and traditional institutions. His argument is that "most tragic episodes of state-initiated social engineering originate in a pernicious combination of four elements" (ibid, 4):

1. Administrative ordering of nature and society

2. High modernist ideology, which "must not be confused with scientific practice. It was fundamentally, as the term "ideology" implies, a faith that borrowed, as it were, the legitimacy of science and technology. It was, accordingly, uncritical, unskeptical, and thus unscientifically optimistic about the possibilities for the comprehensive planning of human settlement and production."

3. Authoritarian states that have and use coercive power over its subjects to implement designs.

4. Weak and submissive civil societies that were not prepared to resist state-enforced plans.

To be fair, while a very important agent of technological change, the modern state is inherently a bureaucratic, centralized and hierarchical structure (Gill 2003). Such qualities present limitations when addressing complex issues such as those of sustainable technological development. However, those qualities of the state have legitimate advantages. Their comprehensive undertaking on development issues, of all types, makes them able to make effective shifts no other institutions in modern societies can make. Those shifts can open doors for many positive frameworks that otherwise had no, or very little, chance. By opening such doors, state policy is capable of making developmental strides that emphasize equity, participation and steady progress. This is one of the reasons why, in today's contemporary contexts, no admired development achievements have occurred under weak states. A strong state, thus, is a pre-requisite for development, as Nyerere put it (1998). Depending on the vision and policy of the state leadership, the outcome can be quite different—i.e. good, bad, or very bad.

## The big picture: strong and weak states

In recent decades, there had been a strongly-asserted, even imposed, argument on developing countries which are struggling economically. The argument is that their way out of the vicious cycle of poverty and debt is through privatization. Privatize your state's assets, they say, make your state agencies and budgets smaller and your private sector bigger; the private sector will meet your development needs if you allow it to grow and take over the economy; etc. This argument had been generally supported, in word and deed, by most of the main players in the international development arena, such as the World Bank, the International Monetary Fund (IMF) and the World Trade Organisation (WTO).

But we have overwhelming evidence to show that weak states cannot be conducive to development, even if other sectors, such as the private sector and civil society, claim to be developed. This rule applies irrespective of the general economic orientation of the country (e.g. market economy, socialist, mixed economy, etc.). The reasons are embedded in historic and logistical-technologic factors, and thus should be addressed as such. That said, however, a necessary remedy to state centralization is needed, especially in the contexts of post-colonial societies where some state regimes tend to be despotic, because centralization has a known tendency against sustainable and equitable development.

In order to clarify what is meant by strong or weak states we have to consider the context of the post-colonial 'nation state', the only officially recognized form of state in our times, and its main features. For that we need to briefly review the history of the formation of this world of post-colonial states. The 'nation state' or modern/contemporary state model is mainly a European model that was 'adopted' by the rest of the world through colonization, and was later emphasized as the only legitimate model through European global hegemony, through organizations such as the United Nations and its predecessor the League of Nations. For that, the 'national identity' attached to this model is actually a geopolitical – i.e., territorial – identity and not necessarily 'national' (from a sense of a nation, which is a homogenous historical, cultural and/or linguistic identity of a people). This post-colonial state is new in human history in terms of the level at which it plays a critical role in the daily lives of every single individual within its boundaries. The mere fact that every single resident in the country must have some type of state-issued document (hence a life-time transaction with the state) attests to that. If that individual does not have such a document they almost do not exist (or are 'illegal'). Other characteristics, such as monopoly over the use of coercive force, the right to international representation of citizens, and the choice of official currency, also attest to that dominance of the state apparatus in our contemporary lives everywhere on the planet. The outstanding characteristics of the post-colonial state have been summarized by An-Na'im (2008, 86-87) as follows:

- "The state is a bureaucratic organization that is centralized, hierarchical and differentiated into separate institutions and organs with specialized functions. But all the institutions of the state operate according to formal rules and a clearly defined hierarchical structure of accountability to central authorities.
- The hierarchical yet interconnected state institutions are distinguished from other kinds of social organizations, like political parties, civil society organizations, and business associations. The scope and functions of the state require its institutions to be distinct from non-state organizations because state officials and organs should regulate non-state entities and may have to adjudicate differences among them.
- The expansive and far reaching domain of the [post-colonial] state—extending now to every aspect of the social, economic, and political life, including the provision of educational, health, and other fields—is far more extensive than any other kind of organization [within its borders].
- To fulfill its multiple functions and roles, the state must [be] the highest authority within its territorial borders. The state must also be the authoritative representative of its citizens and entities within its territory to all entities and actors outside that territorial domain [on most official political and economic aspects].

- For the same reasons just cited, the state must also have monopoly over the legitimate use of force and coercion. This capability is essential for the state to be able to enforce its authority in order to protect its sovereignty, maintain law and order, and regulate and adjudicate disputes."

A 'weak state' is one that has failed to sufficiently embody the characteristics above. A 'strong state', by contrast, is one that has fulfilled those characteristics.[5] Without a strong state modern development efforts are highly unlikely to succeed (although not every strong state is necessarily a developmental state), irrespective of other factors. Weak states cannot achieve overall development, whether they are market economy states that give great power to the private sector or states with more command over their national economies. In Western Europe and North America, and other industrialized economies, states are generally very strong. These states are involved in almost all the major aspects of their residents' lives and livelihoods. They hold the highest right and access to coercive force within their boundaries to enforce clear legislation regarding trade operations, market adjudications and financial transactions. They also have the largest resources to execute their mandates, through detailed and comprehensive taxation systems and property regulations. Yet they can have regimes that promote so-called 'free market' philosophy. It is practically proven that market economies cannot regulate themselves, since they need the legislative and enforcing power of the state to establish the necessary financial and legal conditions to flourish. This is why the biggest markets and consumerist cultures in the world today reside in very strong states. There is ample evidence that economic and human development does not happen, in visible measures, in the post-colonial world, under weak states. Both private businesses and NGOs cannot efficiently undertake the tasks expected of the state in any national development context. As for the local civil society potential, its healthy existence is highly dependent on the foundational legal framework based on principles of constitutional citizenship and participation, something that can only be guaranteed by the state apparatus. One of the prominent figures for development governance in Africa, Julius K. Nyerere (1998), was thus sage in advocating that,

---

(5)     From this it follows that it is expected – and exists – that some strong states are 'stronger' than others, and some weak states are weaker than others. That level of strength, however, still does not necessarily determine the developmental approach of the state authorities.

"[Africa should ignore the call], emanating from the North, for the weakening of the state. Our states are so weak and anemic already that it would almost amount to a crime to weaken them further. We have the duty to strengthen the African states almost in every respect you can think of... In any case dieting and other slimming exercises are appropriate for the opulent who over-eat; but very inappropriate for the emaciated and starving."

## Development, redistribution, and the state

Karl Polanyi (1957) proposed that there are three general forms of integration through which economic processes are instituted in societies. These three forms are:

1. Reciprocity (founded upon patterns of symmetry),
2. Redistribution (founded upon patterns of centricity), and
3. Exchange (founded upon market systems).

Let us focus on the second form of integration, which is redistribution. The pattern of centricity that Polanyi mentions here means that there is a central point (of authority) to which resources flow in, from all the parts of the large social unit, and out again, at the discretion of that central authority. That 'central' authority can be federal, for example, with measures of autonomy to sub-regions, but still central in its overall structure. This political-economic process is epitomized in systems of taxation, public spending, land/resource allocation, regulation of property rights, and chartering priorities of development infrastructure and public services, among others. It is, therefore, the degree and kind of redistributive activity that the state undertakes which makes it either a developmental state or otherwise.

The redistributive role of the state is paramount for developmental concerns, and redistributive decisions that governments make reflect their priorities. A government that is more involved in directing its economic and human development tools towards pre-conceived goals would utilize the state institutions accordingly and make them big agents of development (and technological change). On the other hand, governments that seek to minimize the state's role in industry and trade, with the claim that the market will take care of itself and of the economy, are likely to create the opposite of a developmental state. It might, at best, be what has become known as a nightwatchman, or minimum, state concerned with territorial integrity and safety. From modern history it can be shown that countries that improved their overall development in the last century, comparatively, generally used state-driven approaches.[6] The tools and priorities through which states and markets manipulate economic spheres are quite different from each other; therefore, market forces cannot claim to substitute the role of the state in the economy. Governments that choose to delegate economic development to the market end up tailoring the state's role to that of a guarantor of a legal and financial environment conducive to the private sector's interest, notwithstanding that the state's involvement is never diminished. This is why some writers see attempts at public-private partnership as a double-edged sword—they can be a way of diluting the redistributive role of the state "in favour of a technical, marginal one" such as monitoring and insuring, which is the case with many such partnerships (but not all) (Miraftab 2004), or they can be used to allow the redistributive role of the state to achieve big developmental leaps with equitable results in society (Gilbert 2009).[7]

### Decentralization as a remedy

That said, it is apparent that the very structure of the post-colonial state has inherent problems that challenge principles of genuine participatory democracy (as opposed to mere representative democracy) and socioeconomic self-determination for smaller social units within the larger social aggregate. These same problems pose challenges to technological innovation and talents expressing themselves freely and finding institutional support. These challenges however can only be addressed maturely when the conditions are ripe, and with the right political processes. The ripe conditions require a mature infrastructure—of education, healthcare, constitutional rights, national economic autonomy, etc.—to be attained by strong, developmental states.

The state is a child of the irony of historical dialectic. It transformed human identity from the ethnic group, race and religion, to a simple and objective identity: the geopolitical identity. The state also was historically instrumental to establishing universal human rights

---

(6)    Most of these state-driven approaches were either socialist or embedded-liberalism. We can briefly define embedded liberalism as 'market economy with state guidance and intervention' as a principle (i.e. not only in *de facto* politics). While liberalism in general holds that markets should be as free from state involvement as possible, embedded liberalism considers macroeconomic planning as multi-faceted and requiring of state regulation while markets and trade should be as vibrant and 'free' as conditions allow.

(7)    The example Gilbert gives, from the USA, is what happened in the famous 'New Deal' era in the USA's history when social welfare, land reform and access to education measures were widely adopted and deepened, with national wealth, benefiting more people, without really jeopardizing the general growth of the country's economy.

and international law, along with standardized systems of scientific inquiry, technological advancement, and redistribution of power and wealth. On the other hand, the state had an inherent oppressive side since its very beginnings. Its own existence requires the monopoly of coercive force and criminalizing all behavior that is not conducive to its continuation (even when such behavior is not necessarily criminal under moral criteria). The state also amplified class conflict, helped nourish fascist regimes and ideologies, and suffocated many positive and revolutionary forces of society. In the end, humanity cannot deny the lessons learned, and benefits gained, from the state experience, but it has to be recognized that it cannot move towards its ultimate goal – the realization of full potential and emancipation – without eventually making the state part of history. The universal human community, which was able to take strides forward using the state institution, is also capable of crafting alternative systems that learn from the previous experiences and surpasses them. The road towards these alternatives begins, in our opinion, with decentralized federation.

Why? The state largely feeds on centralized authority. Centralized authority allows the state to execute the will of the central government effectively and, most of the time, swiftly. The power of legislating and executing laws keeps the centralized authority on top of things. Too much power is often entrusted to political figures, who in turn often misuse it. Such hierarchical and all-pervasive power in society should always be called into question, asked to continue to justify itself and be accountable to scrutiny, firmly kept on a leash, and continually have its powers kept to the minimum necessary level.

So, how can we forge socio-political systems that embrace the importance of strong developmental states while also mitigating the negative aspects of centralized authority? The course of historical development of the state in post-colonial societies gives us good insight.

In Africa, for example, most African countries have not witnessed a strong independent modern state yet, and it would be unwise and ahistorical to ask them to bypass that stage. The reason for this can largely be traced to the experience of colonization. The native social development process of Africa was severely interrupted and diverged to serve foreign interests that were directly opposed to the interests of natives (Rodney 1972a). The last century witnessed the withdrawal of these foreign colonizers, but only after they had decided on the borders of the 'new and independent' African states and laid out the main features of the governance structures. Africans were then asked to occupy these structures as is, and continue forward normally, as if things should simply continue on that established trajectory without questioning, with the new members of the international community judged accordingly.

Genuine (i.e. authentic and grounded) social progress does not happen that way however. This is why post-colonial Africa was left to catch up with its own native history, but in a sporadic way dictated by the circumstances. This unique African post-colonial context calls for a special analysis. Amilcar Cabral responded to this call by devising a theory of national liberation – its goals and methods – that highlights the importance of the process of entrenching the modern African state into the native social dialectics of Africa, in order to reignite and unlock the local social forces. This process, Cabral argued (1966), will eventually make reasonable and strategic connection with the native African history that was severely interrupted by colonization. Such a process of decolonization generally aims at what could be called the forging of a native modern African state – not simply a post-colonial state – from the remnants of the colonial legacy. It is very important because genuine development in our contemporary times depends on it.

The state coordinates labour and resources to build national infrastructure (roads, railroads, utilities, sanitation management, etc.) and facilitates the provision of other critical public services (education, healthcare, safety, and quality assurance of circulated material goods, etc.). To envision autonomous communities that would actually do that, and build modern developed material conditions in Africa, without a central coordinating body – i.e. a state institution – continues to look unrealistic, un-tempered by history's lessons.

There is no good reason to conclude that Africans would have not reached some sort of modern nation-state model if it was not for European interruption (although the author is generally not a fan of 'iffy' readings of history, for the sake of the argument the thought is entertained). We look into history, and we see the common law of social evolution that groups of people tend to grow from small/simple social aggregates to big/complex social aggregates. This applies to the histories of the nations of all continents. So nationhood, and indeed 'region-hood' (i.e. political and socio-economic consolidation of multiple nations under a common vision and structure) are not surprising cases in that trend. Indeed, Africa itself saw some sort of early pre-colonial versions of those trends. We now see Europe, the continent with a history of internal wars of larger magnitude than any other continent on Earth, is now leading the rally for continental unity.

As far as technological development goes, with all things considered, there are good lessons to learn from the currently impressive technological strides of two countries that were both poor and developing ones a few decades ago—China and India. These two countries now have big space programs with big achievements in space technology, and that by itself counts for something. The two main lessons from that side of the China/India story are, firstly, that modern technological transformation, as a systemic process, is replicable from one nation to another, mainly because it is essentially a human feat of learning and improving under conducive conditions (as elaborated in the previous two chapters). It is not magic and it is not a property of one nation alone in history. Secondly, China and India proved, to a larger extent, that a nation can achieve net material progress to impressive degrees without having to destroy other nations and monopolize their resources by force and brutal violence; i.e. without military and political colonization.

If we observe further how social aggregates tend to grow from small/simple to big/complex, we also discover that when social aggregates grow more complex, a type of centralized political management emerges. This is also witnessed in the known histories of all continents. The imagination of the role of centralized authority is not necessarily a work of the wicked and despotic, but actually a work of objective, universal social history. Africa is no exception. However, the African model of the nation state does not have to be – and should not be – a copy of the European model. The African state can be different. That is where a revolution can happen, when the people reclaim and re-design the state apparatus to work for their interests instead of elites. It is here that the measures of decentralization are an important instrument: because it is more capable of, and conducive to, carrying and responding to the pulse of the masses effectively. The contemporary answer, for Africa, I think, lies in the decentralized model of the state, which benefits from the state structure while mitigating its negative impacts.

This model is broadly described as the federation model (with many degrees of federation). Federation can be defined as the political system formed by the union of several self-governing units (e.g. provinces, localities, etc.). In a democratic federation, the central authority shares political powers with the federal members, in a way that balances and decentralizes political and economic control to the benefit of citizens. The federation model is difficult but necessary. Since excessive centralization of political authority is not conducive to equitable and sustainable, people-oriented development, a remedy to excessive centralization should be sought in federation. Federation, in theory, is an anti-thesis to centralized state domination, but not a promoter of 'weakening the state'. Decentralization is also not only reflected in political structure and decision making, but in development models and cultural living as well. A decentralized development model will give more priority to economic investment in regional schemes for local agriculture, energy harvesting, and water supply, for example. These schemes are more inclined to invest in native management and more environmentally sound than state wide similar schemes controlled by the central authority. On the cultural front, and with the reality of cultural diversity in most post-colonial countries, a decentralized model would more likely help different groups express themselves, through their cultural life, with more agency, and with less imposition of a state-sponsored cultural identity. That in turn is more likely to catalyze genuine coexistence and the peaceful reaching of consensus on an inclusive national identity. A strong state, in the decentralized federal model, compliments that trend by coordinating efforts and channeling resources more effectively than a weak state would.

While the strong-but-decentralized state model is not necessarily 'the answer' to the problem of development in post-colonial societies, it is a critical part of that answer. One cannot dismiss the role of the state as a critically influential actor in technological change, but the degree and type of that influence cannot be taken for granted, and needs scrutiny with clear orientation in policy and resource deployment and redistribution.

## Agents in the market[8]

Private sector agents that work in the market (e.g. private firms, technical consultancies, etc.), under conducive conditions can champion significant technological R&D activities that can lead to improved products, processes and services.[9] However, there is a difference between the market, on the one hand, and market economy on the other hand. What applies to the market economy, as an economic system, does not

---

(8)     Based on "The market is alright, but not the market economy." August 18, 2016 on *Pambazuka News*, Issue 788: http://www.pambazuka.org/economics/market-alright-not-market-economy

(9)     See for example, Arthur 1989; Monden 1993; Rogers 2003.

necessarily apply to the market itself as an economic phenomenon (a single institution).

On the one hand, the market is an arena for trading commodities, where various groups and individuals express their preferences among comparative, complimentary and exchangeable goods that have relative utilities. The dynamics of the market mechanism include competition, promotion, mutual learning, and balances of the utility-quality-price triangle, but only with respect to commodities (including technological products and services), not with respect to the whole economic cycle and industries in society. Markets have been known throughout history to do well, under conducive conditions, in circulating goods widely, improving standards of quality, responding to consumer feedback, and promoting innovation in fair competition.

On the other hand, the market economy is larger than the market alone, and hence it means something largely different. The market economy is an economy that puts the market - defined above - in front of the steering wheel of the entire economic life of society: production, relations of production, redistribution, and development and provision of services. The main idea is based on the postulate that market mechanisms can be self-regulating – i.e. self-monitoring, self-correcting and self-sustaining – and therefore should not be regulated by social or political authorities. Moreover, it is claimed that as they become self-regulating they eventually guide the entire economy in more objective and effective ways that respond and adjust to the real world of supply and demand. Markets naturally support 'rational choice', it is argued. Consequently, however, we find that in market economies, decision making in economic matters is generally not a public affair (hence not democratic) and it particularly rests in the hands of a consortium: an alliance between those who control the flow of commodities in national/regional markets and those who wear the hats of policymakers within those same national/regional boundaries (because, after all, it turns out, markets need to have the policies and legal protection in place that allow them to self-regulate). Sometimes we find that a vague voice for scattered, disorganized and underrepresented consumers is added to that consortium (or allegedly so). The underlying assumption is that this consortium simply responds to the objective trends that show themselves in supply and demand. This assumption is vulnerable to questions however.

Karl Polanyi (1944) contended that market economies eventually create market societies, market politics, and overall market cultures.

It happens because economic life, in any given society, permeates through all of the above (social life, politics and culture), and the market economy needs to control all of them in order to control economic life. Polanyi explained that the market achieves control through a number of major alterations (or transformations), the pinnacle of which is the act of taking things which are essentially not commodities and treating them as commodities. This act is the biggest hat-trick the *laissez-faire* doctrine ever pulled—creating 'fictitious commodities'. Polanyi named these as: land, labour, and money—also known as the elements of industry. Fictitious commodities are called so because they do not meet the original definition of a commodity, but the market economy pulled all the strings it could pull – theoretical, legislative and administrative – to see that they were viewed and treated as commodities. The definition of a commodity is a good that was created or devised for the purpose of trade/sale in the market. The three fictitious commodities, mentioned above, do not meet this definition. Let us take a quick look at each one of them:

Land is not a commodity because it is another word for nature. Humans did not create or devise land; the other way around sounds closer to the truth. When the value of land and its utility is determined by the market it is commodified (i.e. made into a commodity) irrespective of the many complexities that entangle humans and their natural and built environment which cannot be reduced to simple 'property value'. Generally the market price of land is called 'rent', and in its absolute versions leads to the disintegration of nature and violations of environmental balances and long-standing socio-ecological relations. The alarming ecological degradation in the planet today is largely related to practices of land commodification normalized by the market economy.

As for labour, it is a form of human activity, an extension of humans, combining their physical, intellectual and psychological effort and springing from skills and knowledge earned through learning and experience. Labour cannot be a commodity unless you take a human being and own/control their effort in a particular time and place. Having agreed to it at legal levels does not make it less problematic at the philosophical level. Commodifying labour has an element of commodifying a person's existence in a particular context, for which the price is paid and determined by others with power. The price of labour in the market is called wage, salary, etc.

Finally, money was created by human societies as a medium of exchanging commodities, but not as a commodity itself. Money took many forms (of currency) throughout history, from salt to minerals to tender paper, yet always as a way to exchange various commodities in the market using a standard method of payment, but not as a commodity itself. The idea of 'buying money with another money' seems automatically ludicrous to our sensibilities for that reason. However the market economy succeeded in making a huge 'market' for the activity of selling and buying money across the globe—through interest rates, debt financing, derivatives markets, foreign exchange markets, and other related financial transactions. In such markets, money can literally get you more money without any productive work (value addition) or genuine commodity exchange taking place. Folks can get richer for no other reason than that they are already rich.

**What it means for development**

From the above it should be relatively easy to see that the commodification of land, labour and money eventually leads to society and nature being under the control of the market, which is a recipe for disaster. Yet it is a recipe that has been normalized under the *laissez-faire* doctrine to the point that we rarely, if ever, ponder it in our daily lives. It should also be easy to see that the market economy is quite different from the mere institution of the market. Markets can exist and thrive without market economies, as many did earlier in pre-capitalist and pre-colonial history.[10] It can also be reasonably imagined that if the market economy existed in its pure manifestation (i.e. totally loyal to the theory), in any country, it would soon be widely rejected for its monstrosity towards humans and nature under the vague goal of 'economic growth' (i.e. assuming that economic wealth is an end in itself, separate from the social ideas and aspirations of prosperity and perceptions of what constitutes a good quality of life). Instead, the countries that today adopt the philosophy of the market economy in theory ameliorate its raw effects in practice with many contrary measures in order to reduce its zeal (as explained earlier). Such measures come in the form of packages of social safety net programs, free public services, labour rights – just enough to not

---

(10)     In early empires and chiefdoms, indeed it happened frequently that there were very big and vibrant markets that did not, however, control the macro-economies and lives of their empires/chiefdoms or all terms of trade. For more details on such cases, see Polanyi, K, Arensberg, C. M and Pearson, H. W (eds.) *Trade and market in early empires: Economies in history and theory*. Glencoe, IL: The Free Press.

look bluntly inhumane – and some minimal environmental protection laws (that can be broken if a corporation can afford to pay the fines in a business-as-usual manner), etc. These countries do in practice what they do not support in theory, and so they do it half-heartedly. That is largely why many, including the author, find that it is more coherent to reject the market economy, in clear terms, without rejecting markets. Rejecting the market economy essentially amounts to rejecting the commodification of land, labour and money – i.e. to decommodify them – and to carry out economic decision making in society in a democratic and inclusive manner, especially including the creators of value in society: the workers. Rejecting the market economy means running the economy as an adjunct to society, not the other way around. This path can be followed in several ways, and in various accents depending on the context of the society in question (such as agrarian or industrialized economies, low or high income, etc.), once we dispose of the idea of adhering to the market economy framework. This path can be followed though a package of conscious, bold and unapologetic policies for the economy, such as:

- Activating and normalizing the redistributive role of the state and its institutions for socioeconomic equity and growth.
- Cementing strong and comprehensive labour rights (including organization and representation in decision making).
- Cementing strong and comprehensive laws and procedures for ecological balance, agrarian reform, land use and tenure, etc.
- Vividly discussing and implementing policies on minimum and maximum purchasing powers to generalize in society.[11]
- Invoking, encouraging and supporting the development and proliferation of cooperatives (in industrial, agricultural and service sectors) through the framework of cooperative economics.
- All the above while also allowing markets to flourish for genuine commodities (including technological products and services),

---

(11) Adam Smith, proclaimed father of capitalism, warned against allowing the existence extremely rich individuals in society. He explained that this was clear evidence of a failed economy, since a successful economy should enhance the livelihoods of the entire population. Having extremely rich individuals technically requires (and implies) that we have quite poor people in that society as well, which means that the economy at large is not balanced and free, i.e. not a success. Extremely rich individuals, he also contented, have the tendency to monopolize industries and therefore hinder fair competition in a 'free market' arena. Putting a general and reasonable cap on how much wealth any individual or conglomerate can accumulate in society is no farfetched idea and is about protecting society from economic tyrannies. This is the argument of an early capitalist who did not live to see the consequences of normalizing the market economy idea.

through regulating them for quality, applying customs on selected import goods, monitoring logistics, consumer protection and fair competition.

To summarize this approach: nowadays the market economy creates a double-fold falsehood: it makes society a subspace of the economy, and the economy dominated by the market. We propose to correct the falsehood by making the market a subspace of the economy, and the economy subordinate to society.

It could even be reasonably argued that, once freed from a burden they were not meant to bear, markets may even become better and healthier for genuine commodities and their producers and consumers. They could also become vibrant grounds for innovative industries – such as small and medium-sized enterprises (SMEs) – to unlock their potential as agents of technological change.

## General orientations

In this section I discuss main orientations that inform the approach of this book's advocacy for technological autonomy in developing societies. The orientations are conclusions for policy thinking that help agents of technological change address it more effectively, learned from the study and analysis of historical experiences as conveyed by the author through general recommendations.

### R&D priorities: 'technology & science' or 'science & technology'?

The renowned biochemist and philosopher, Lawrence Joseph Henderson, is attributed with once saying, "Science owes more to the steam engine than the steam engine owes to science." It has been explained that what he referred to is that technological breakthroughs are often the ones that usher in scientific breakthroughs, more so than the other way around. This is not always the case, of course, but has been a consistent trend in most of human history. The industrial revolution was no exception. For instance, the engineering work of the steam engine ushered in new attempts to understanding heat transfer and thermal energy, leading to discovering the laws of thermodynamics. That is the argument, and while it is highly debatable in today's intricate integration of cutting-edge high technology and scientific frontiers, it nonetheless is of appreciable relevance in the context of developing societies.

As mentioned earlier, priorities become important in the face of conditions of scarcity of resources. In R&D efforts that are publicly funded (or funded through non-public channels but in-sync with public/national plans and priorities), there is a need to decide where and when to allocate resources, as well as the size of such resources. In the general language of technological change and development, we often speak of 'science and technology' as a habit. But with that language comes an underlying assumption—that science and technology usually go together, and often science is a precursor and prerequisite of technology. Yet the reality of technological change is not necessarily in agreement with that underlying assumption.

A policy brief, titled "The Utility Value of Research and Development (R&D): Where does Tanzania Stand?" published by the Science, Technology, and Innovation Policy Research Organization (STIPRO), Tanzania, addresses this particular issue in an articulate manner (2010):

> "While the positive relationship between investment in R&D and social and economic development cannot be denied (most countries that commit a remarkable proportion of their GDP [gross domestic product] to R&D are the rich ones), the direction of causality is very much questionable. Does investment in R&D bring socio and economic development or do richer countries tend to spend more on R&D?"

Essentially, the importance of R&D investments to national socioeconomic development is undeniable, but not any R&D. There has to be intentional and strategic prioritization and allocation of resources, and that has to be informed by the realities of the country/region in question, not a foreign reality or a future aspiration. Therefore, it is rather a question of how much R&D investments are enough for certain contexts, and what their priorities should be. STIPRO's policy brief makes a very important observation here: that the contemporary history of technological change in national contexts show that there are particularly dominant trends among countries.

> "I. Those countries where innovations to a large extent depend on imitation and minor technological improvement, where formal R&D is normally not required: Innovation is achieved through tinkering or learning by doing; and where R&D is used, it is to assist in the tinkering; and in most cases has to be close to production, e.g R&D units in companies. The example of Japan is instructive here. Japan for instance started industrialization through import of foreign technology, integrated this into R&D and production departments. The Japanese R&D during this era, termed as catching up period (1945-1972) was largely on adaptive technology.

II. Those countries that are capable of generating and putting in practice new technologies (products and processes). Here R&D is normally a part. Investment in R&D by both the country and companies is therefore a must."

These trends should be able to inform us on priorities, and that, "technology should therefore be put first rather than science; because without technology science is meaningless as far as socioeconomic development is concerned – the abbreviation S&T [science and technology] in this case would have been T&S" (STIPRO 2010). The mental attitude that could result from the switching of these two words might be easily underestimated. Yet an important point is made: on the path of national technological change priority should go to harnessing and enhancing technological capabilities, tinkering and technical innovation. Modern scientific R&D endeavours are indeed important, but they often need a requisite level of technological preparedness to take place, not the other way around. This is more pronounced in the context of developing societies. A proper orientation on the conceptual side is important. The operational side, which follows, will be informed and guided by the conceptual orientation.

If we orientate in a T&S manner, we would, for example, prioritize 'applied research' over 'basic research', as commonly called, (yet without dismissing basic research on important and exceptionally curious frontiers). The difference between the two, I may simplify, is that basic research explores the frontiers of 'what we know', while applied research explores the frontiers of 'what we can do', and the latter is more contextual, result-oriented and closer to the T&S approach. We would also give more value to reverse engineering, which is the process of learning the composition of a technological product in a practical way, by deconstructing its parts and understanding the details of its composition, in order to reproduce it locally. By 'locally' we mean at the level of the R&D place of operation (whether a laboratory, workshop, firm, etc.), which may eventually extend to a national or regional level. We know from historical records that several countries in the second half of the twentieth century achieved big leaps towards industrialization through systemic reverse engineering efforts. Notably, this happened with the Japanese automotive industry as it reverse engineered American and European cars, and it also happened with other technology sectors in countries such as South Korea, Brazil, India and China. Malaysia, as another example, was able to acquire the national technological capabilities to become an automotive manufacturer through a process of learning and reverse engineering Japanese automotive technology[12], in a period between 2 to 3 decades. The benefits of reverse engineering mainly rest on the proven result of the transfer of both the technology products and the technological expertise, to local hands and industries, and then it opens the doors for local modification and innovation.

Industrialization often comes in the form of growing raw material processing activities instead of exporting them as such and importing finished goods. The United Nations Sustainable Development Goals express that fluently. Under goal 9, which concerns the promotion of sustainable industrialization and building resilient infrastructure, one of the aiding trends mentioned is that, "In developing countries, barely 30 per cent of agricultural production undergoes industrial processing. In high-income countries, 98 per cent is processed." Additionally, it is mentioned that, "Small and medium-sized enterprises that engage in industrial processing and manufacturing are the most critical for the early stages of industrialization and are typically the largest job creators. They make up over 90 per cent of business worldwide and account for between 50-60 per cent of employment."[13] It is fairly common knowledge among those who study technological development that, "Industry—including manufacturing, agro-industry, and tradable services—is a high productivity sector that has the potential to absorb large numbers of modestly skilled workers, contribute to accelerated poverty reduction, and diversify the economy" (Page 2016, 1). The positive connection between technological, human and economic development is established in more than one way.

Existing infrastructure and institutions could be used differently, after re-orientation. For example, a number of developing countries (some of which are now sufficiently industrialized countries or beyond) used R&D parastatals – otherwise called public technology intermediaries, or PTIs – or industrial support organizations to push and promote technology localization. These R&D parastatals are members of a myriad of parastatals that are typically semi-independent organizations,

---

(12)    Such is the summary of the story of the Malaysian automobile brand 'Proton', which grew from partnership between a Malaysian parastatal and the Japanese car manufacturer Mitsubishi. In the 2000's Proton made Malaysia the 11th country in the world with the national capability to design and produce cars from the ground up.

(13)    United Nations Sustainable Development Goals. "Goal 9: Build resilient infrastructure, promote sustainable industrialization and foster innovation." Retrieved on May 15, 2017 from the official website: http://www.un.org/sustainabledevelopment/sustainable-development-goals/

with distinct legal structures and procedures from state agencies. The state is sometimes the owner and sometimes the main shareholder.

Many parastatals in developing countries were established between the late 1960s and early 1980s. The rationale behind their creation, according to the philosophy of authorities at the time, was for the state to guide the development of particular industries and key technology-based businesses that needed stimulation or were too important to be left to the private sector alone. In Tanzania, for example, parastatals stretched along financial, agriculture, infrastructure, manufacturing and service industries, with a number of them still surviving to date. Their broad and varied experiences have been widely studied.[14]

Although parastatals are common in many countries around the world, they usually happen to be commercial parastatals, i.e. engaged in commercial activities as state owned (or shared) enterprises. R&D parastatals, on the other hand, are not as common or visible. Some of them prove to be key contributors to their countries' industrialization. For example, when Taiwan was entering an era of industrialization in the early 1970s, its now-famous R&D parastatal, Industrial Technology Research Institute (ITRI) "played a vital role in Taiwan's economic growth as it shifted from a labor-intensive industry into a value-added, technology-driven one." (ITRI nd.). ITRI combined technological innovation (patented) and enterprise incubation to conceive and diffuse key products and systems in Taiwanese society (Ash et al. 2006). On the other hand, Vietnamese R&D parastatals are credited for their significant contribution to raising Vietnam's technological profile as it grew from a low-income to a middle-income country (Wolff 1999). Many R&D parastatals were established in the 1970s and 80s in developing countries, but some began later in the 90s. Typically, R&D parastatals promote innovation through technical assistance, design and incubating ideas that respond to local technical needs. Typically, they are co-located with researchers, engineers and technologists sharing offices, libraries and machine workshops. Currently, in some developing countries, they may still exist but are under-funded and operate within institutional constraints that limit their potential in contributing to technology localization. Still they have clear potential. In re-orienting technological research towards a 'T&S' approach, national policies could make use of such organizations and start the re-orientation with them. They are also very viable candidates for places of extensive reverse engineering

---

(14)    See Loxley and Saul 1975; Coulson 1982; World Bank 1988.

activities. As a matter of fact, many of them are no strangers to reverse engineering, but they could have more of it.

But a policy of re-orientation and re-prioritization generally extends to many areas of learning and doing—education, research support, national development flagship projects, priority sectors, etc.

## Regional technological cooperation

It is economically cheaper, logistically more efficient, politically more stabilizing, and commercially lucrative to integrate industries regionally (i.e. between countries in the same region). It is more so with developing countries (although industrialized countries do it even more often and for the same reasons). It is also intuitive that regional technological autonomy is a reasonable precedent to national technological autonomy.

J. K. Nyerere (2011) articulated this in 1980 while speaking to fellow leaders of African countries, "We cannot have fifty iron and steel complexes in Africa at the present time. All of them would run at a heavy loss which we cannot afford....We need to coordinate our industrial strategies, to exchange technical information, and to work out some means of concentrating those industries which demand very heavy investment and large market."[15] When the regional supply chain of products and services is strengthened the regional market flourishes, and that economic stimulation not only stimulates more technological autonomy but also breeds more confidence in regional capacities to meet greater demand. Regional industrial cooperation also stimulates stronger regional transportation and communication networks, regional tourism, regional food and energy security. Eventually, if it perseveres, all that leads to recognition of the importance of continuing to develop endogenous capacities while also acting in greater harmony with each other when addressing the outside global market.

The many benefits of regional industrial cooperation are no secret to most actors and observers in the global political economy scene. The challenges that face the initial stages are also no secret. Nowadays we have examples of regions that attempted it and achieved relatively big successes, on various scales, and regions that discontinued from that path at some point because they were not able to handle the internal and external challenges of continuation and to keep focus on the big picture and net benefits. Lessons can be learned from both.

---

(15)   Nyerere's speech at the Extraordinary OAU (Organization for African Unity, the predecessor of the African Union) Economic Summit Meeting. Lagos, Nigeria, April 28, 1980. See Nyerere 2011.

## Emerging agents of change?

In developing societies, there are technological development challenges that neither the public sector nor the private sector have been successful in meeting. Under such conditions, new agents of technological change sometimes emerge to fill gaps left by the old, or large, or conventional agents. The door should remain open for such emerging agents, as they may strive to play small or big, temporary or permanent roles in advancing technological autonomy. For example, the author recently investigated the potential role of social enterprises as agents of technology localization in East Africa, and the results turned out to be affirmative, with some impressive cases from modern-day Tanzania (Sheikheldin 2017). It is no surprise, however, because the arena of technological autonomy demands and accepts the attention of various actors and models of organization in every society.

Chapter Four

# Influences on Technological Affairs: Politics, Cultures and Ecologies

*"When everything is connected to everything else, for better or for worse, everything matters."*

–Bruce Mau

We have established that there is more to technology than simply 'technology'. There are many institutional factors at play, and many inter-influences that affect technical procedures and results. This chapter will explore key influences on the technological change process, and key topics to reflect upon when we strategize and operationalize liberation technology. This chapter builds on the points and perspectives discussed in the previous ones and explores points of importance to the interface of technological change, autonomy and development.

## From simplicity to megatechnics: a philosophical view

There is a deep relation between humans and their technologies. The relation stretches back to prehistoric times and has always been part of our social and cognitive makeup. It is therefore wise to contemplate deeper that connection.

Mumford (1967 and 1970) traces two essential prerequisites to human technology in histories of societies: namely language and social structures (with the former preceding the latter). The cumulative nature of technological capacities in human societies would not have been so without the ability to communicate thoughts, experiences and aspirations among humans. Furthermore, the control of human behavior and human body[1] – which is the task of social structures – was a prerequisite to establishing the early tool-making faculties of humans, and the accompanying faculty of translating novel images from dreams into relative material embodiments. In modern societies, Mumford argues, the convergence of science, technology and political power tends to render very large systems – "megatechnics" – that overwhelmingly control human expression (outward and inward) and not always in good or favourable ways (such as the example of the military industrial complex). The entanglement of humans in megatechnics complicates our relationship with technology, in the modern age, in unprecedented ways. This same modern reality was articulated by Ursula Franklin when she said that "technology has built the house in which we all live."

Aunger (2010, 762) further addresses the philosophical/historical concern about technology, humans and society. Technology-institutional dynamics are indeed not just modern stories:

> "Technology…appears to be a central driver of events we care very much about – human evolution and history – yet we have very poor understanding of how it evolves. To understand the human condition, we must be able to explain how human technology has become increasingly complex, and increasingly central to modern life, while the technology of other species remains mired in a much more primitive state."

Aunger focuses on the task of tracing the evolution of technology as it accompanied the evolution of species themselves, till it reached human beings and continued to evolve within their societies. One thing he highlights is the technological base we share with other earthlings: e.g. 'dwelling production' in the form of, for instance, bird nests, 'trap production' such as spider web-traps, and 'stigmergy' (or group colonies) in the form of beehives and ant nests, etc. The point Aunger highlights is that we have to view technology as an essential component of human evolution itself, part and parcel thereof, instead of as a mere by-product of that evolution. A number of observations can be highlighted from this examination:

---

(1)  Mumford refers to the human body as the first machine that all humans experience and seek to understand and utilize. An interesting perspective but this would pose problems to the 'artefact' component of technology we discussed earlier. For practical purposes, and to keep consistency of terminology within the book, this paper would not entertain the idea that the human body fulfills the definition of a machine.

1. While technological evolution can be traced across the wide spectrum of species, it is also noted that generally all species, save humans, are locked in certain levels of technological evolution. Change in technology – upward or downward – accompanies a change in the species themselves.

2. At the human level, we find a number of technological transformations that occur over periods of history, with dramatic effects on societies and epistemology. One example is the making of tools that can only function when used together, but not separately (like bows and arrows). Another example is the transformation from making tools to making tools that are only used to supplement other tools (i.e. machines that only serve to compliment other machines or processes, but do not have direct human use, such as animal-drawn ploughs, or bridges). This 'separation of making from using' was a profound transformation in our relation with technology, as individuals and as societies.

3. The separation of making from using paved the way for another profound transformation in human relation with technology, namely exchange. This also resulted in human beings making things solely for the purpose of trade rather than for their own use (direct or indirect). This not only furthered the separation between making and using, but also separated makers from users. Thus was born the technologically-supported division of labour, or 'specialization in production'. Division of labour, as we know, had become the founding social principle upon which complex social structures (including political and economic institutions) grew to dominate human lives.

4. With the rise of modern technology emerged technological systems. "In systems, heterogonous artefacts become developed in an organisation or 'industry' such as telecommunications [and power generation, etc.]" (Aunger 2010, 775). Technological systems transform human lives, once more, in unprecedented ways (similar to what Mumford termed 'megatechnics'). This time even human-social organizations and networks cannot fully control technology, its growth – vertically (in level of technical sophistication) or horizontally (in social diffusion) – or its consequences.[2]

---

(2)    In fact, some circles use the term 'technological autonomy' quite differently from how this book uses it, to refer to the point above: i.e. how modern technology appears to continue to become more independent of human direct control, and the implications of this trend.

Such dynamic growth is sometimes further dramatized by particular sectors of technology. In our age, the astronomically accelerated growth of ICTs is a wonder to observe, for instance. Thus, we live in the age of megatechnics, and the most pervasive characteristic of this age is the omnipresence of intricate and complex technologies, forming technosocial systems, and our clear dependence on that omnipresence, as societies, and as individuals. We reached this age through a long historical process of shaping our technologies and allowing them to shape us. And as we look into the future, this strong connection is more likely to get stronger and more encompassing. Such phenomena would make any keen mind contemplate its philosophical and existential implications, in addition to the social, political and economic implications, at length.

## Where context matters

In the introduction of the book I spoke about the two extreme examples of technology adoption: ICTs and agricultural technology sectors. While, for example in Africa, the ICT sector has experienced rapid adoption across the continent, the agricultural sector has been facing virtual stagnation with slow rates of mechanization adoption and minimal increases in productivity, despite decades of diffusion efforts. What could explain the stark differences between these two stories of technology adoption in the same social-historical context? We begin to see, from such examples, that variables and agents of technological change in the same societies may express themselves differently according to the technology sector and a society's existing priorities.

For example, agricultural mechanization undertaken by the USA and Japan took different paths according to their contexts. Japan invested more in "biological, yield-raising technology, much of it supported by heavy investment in irrigation," due to Japan's limited land area and abundant labour availability. Investment in agricultural machinery was not found to be as useful as investment in biological technology to optimize and increase yield. On the other hand, and with the abundance of land and scarcity of labour, the USA found it useful to invest in mechanical technology even before 1880. Although there was interest in yield-increasing biological technology, which was adopted in the 1870s, "it did not introduce big increases in yield until about 1930, well after the major land frontiers had been closed and mechanization was far advanced." The conclusion is that agricultural mechanization is likely to be localized in different ways (i.e. mechanical or agronomical) due to the combination of land and labour factors of production.

Technological policy agenda can be set to distinguish how the process towards technological autonomy unfolds with regards to the peculiarities of each technology sector. For example, while agricultural mechanization in Africa needs to find a way to either replace or coexist with customary agricultural technologies, ICTs do not have to deal with that complexity as they diffuse into most developing societies. Using the concept of technological embeddedness (discussed in chapter 1) we could argue that ICTs do not face significant resistance in adoption rates compared to agricultural mechanization, because ICTs are not replacing any technology sector that already has its customary/traditional versions embedded in local institutions. The process of replacing or absorbing existing technologies that perform the same functions in that context could be a big reason for the slow adoption rate of agricultural technologies.

## Recognizing and engaging early adopters

In the literature of diffusion of innovations, Rogers (2003) tries to illuminate the process that takes place between the time when an innovation comes into existence and the time it becomes reasonably (or widely) adopted by social groups. He identifies four main elements in this process of innovation diffusion: 1) the innovation, 2) communication channels, 3) time, and 4) the social system. He also identifies the change agents of this process as "an individual [or organization] who influences clients' innovation-decisions in a direction deemed desirable by a change agency." The adopters of innovations are categorized, according to their level and time of involvement in the diffusion process, as a) innovators, b) early adopters, c) early majority, d) late majority and e) laggards. All these elements and agents are used to build models to identify and measure variables determining the rate of adoption. This model gives significant weight to the dynamic relation between technology and institutions (or 'innovations' and 'social systems'). The model explains success and failure of cases of innovation – or their rate of diffusion – according to how adopters adopt or refuse the innovation over time. Communication channels are vital in this process, as change agents utilize them to diffuse the innovation into a particular social system. Case studies for Rogers' model range from simple water purification techniques in a Peruvian village and solar energy in the Dominican

Republic, to agricultural technologies in Iowa, in the USA, and the birth of the laptop in Toshiba R&D labs.

Early adopters, in this model, are defined as the "trendsetters" and leaders of change and adoption of new ways of doing things. Early adopters tend to be relatively more capable economically, in comparison to others in the social group (i.e. capable enough to take risks, but not necessarily 'rich'), and are more informed and connected; hence well-respected in their communities–enough to be taken seriously when they promote something new. This is in contrast to innovators, who often are looked upon as eccentric and too nonconformist to influence a significant number of their peers. Research on early adopters has been increasing in the fields of development, education and business due to the critical and decisive role of early adopters in the innovation adoption process.[3] The role of early adopters in each society can be explored in ways that can guide strategies of diffusion where a process of technology localization is taking place.

### Learning and changing: technology and skills

The process of gaining new and advanced technological skills, and applying those skills to render visible results is often a time-consuming process. It is intuitive that learning curves play a role in the improvement of the utilization of new technologies, as well as the improvement of these technologies themselves (i.e. when people get better at utilizing the technology they also sometimes get better at realizing how they can improve it). Yet there is still a need to articulate this intuition and support it with evidence, in order to make it a legitimate and verified observation. If we consider the steam engine a primary invention, for example, we can then observe many 'secondary inventions' that improved the steam engine over a considerable period of time:

> "Although we find it a convenient verbal shorthand to speak of the "displacement" of one technique by another, the historical process is often one of a series of smaller and highly tentative steps....Enos has studied the length of the lag between invention and innovation for 35 important innovations and has subjected his results to statistical analysis. In his sample the arithmetic mean for the interval between invention and innovation is 13.6 years. Enos defines the date of an invention as "the earliest conception of the product in substantially its commercial form" and dates an innovation from "the first commercial application or sale." (Rosenberg 1972, 8).

---

(3)     See for example Brint et al. 2011; Worthington et al. 2011; Huh and Kim 2008.

Rosenberg (1980) also mentions that, with the institutional separation between modern science and R&D circles, on the one hand, and machinery making circles, on the other hand, conceptual solutions take long periods to translate into working machinery forms. The institutional separation plays a part, but so do factors such as economic feasibility of commercial production, connection with willing investors (public or private) and even the physical infrastructure availability to make the proposed concept operational.

All the above factors interplay in contexts of technological development and influence any such process greatly.

## Technology, culture and social systems

While it is generally difficult to find a comprehensive, agreed-upon, definition of culture, if asked for a concise definition to accompany the approach to culture in this study, in the context of developing societies and technological change, my answer would be Steve Biko's (1978) definition: "A culture is essentially the society's [own] composite answer to the varied problems of life. We are experiencing new problems every day, and whatever we do adds to the richness of our cultural heritage." And because culture is a composite answer, it is comprehensive, non-departmentalized, and inherently dynamic, because every time the conditions of life change on one aspect or another, a culture changes in response. While this definition by Biko highlights the importance of culture it also brings forth critical attention to those who think of culture as a static phenomenon to which loyalty can only be shown by 'preserving' the status quo (of culture) and opposing change.

It should be intuitive that all approaches to technological change and development, at any level, need to pay careful attention to the perception of the technology users (and prospective users if the technology is yet to be realized), who in turn would automatically bring their cultural institutions with them to the process. Moreover, in a local context, the people involved in these approaches to promoting technological change, at all levels, would not be divorced from the technology users, if they were not themselves members of the same user category.

The effect of culture on technology, as documented by the relevant literature,[4] seems to only be highlighted when the case of technology diffusion or transfer fails for whatever reasons, among which may be the

---

(4)    See for example Rogers 2003; Adeel et al. 2008; Dengu et al. 2006; Lucena et al. 2010; Haug 1992.

unsuitability of the technology in mention to the culture of the society/ community in mention. As mentioned earlier, in chapter one, when cases of technological change are reasonably successful[5] it is not often highlighted how the engagement of the local culture was key to making that technological change process a success. This might be explained by the possibility that, when the technology users are in considerable agreement with the technological change process, the cultural suitability proves to be embedded in that agreement without any particular need to highlight this. People's cultural perspectives, after all, go with them wherever they go, and influence their acting in the world in the various ways observed. What this explanation then suggests is that paying attention to people's cultural institutions usually comes along with the genuine participation of technology users in the technological change process. Moreover, it could be argued that this is actually the best way, if not the only way, to giving sufficient attention to cultural influence on technology: involving the targeted communities in the decision making and project implementation from the beginning. Even if the technology in question requires a level of cultural change to be accommodated, that decision can be made by the technology users about their own culture. Culture, after all, is a dynamic existence, as we said, and change is part of it—so long as that change is an expression of agency rather than being enforced by external forces with little understanding of, and identification with, that same culture.

## Language in technology learning and communicating

One major aspect of culture is language. Language also has quite a peculiar relationship with technology in the context of development. Since modern science and technology had mostly been produced in European languages, which had also been spread throughout the world through colonization and trans-continental slave trade, we are today left with a reality of diminishing native languages in developing countries. The ones that are still alive are largely and rapidly losing their relevance to people's daily lives, and therefore dying, because we now live in societies that give more weight to mastering international languages of technology, trade and global communication. Yet shifting from one language to another is no minor thing; especially if the other language is becoming adopted as a main language, not simply a second

---

(5)    See for example Lekoko et al. 2012; Fidiel 2005; Gulrajani 2006; Gilbert 2009; Al-Ghafri 2008.

language. As Frantz Fanon expressed in his book *The Wretched of the Earth,* an anti-colonial treatise written in his colonial language, French, "to speak....means above all to assume a culture, to support the weight of a civilization." This book is also written in the English language, by an African author, and guaranteed to be accessible to more Africans and non-Africans in that form than if written in Swahili, or any other native African language, and other parts of the world where the populations are not 'English'—one obvious legacy of colonization. It is not so much the learning and integration of foreign languages that poses a problem for developing societies; it is rather the perpetual displacement of their native languages by these foreign languages.

Some scholars on development history maintain that it is critical for national development to strengthen and improve native languages and keep them from disappearing from people's lives in trade and technology. Kwesi Kwaa Prah, an African linguist and development researcher, proposes that "No country can make progress on the basis of a borrowed language".[6] Prah highlights that native languages are always important. Just as Icelandic and Italian are important to their people, so are African native languages that have been forcefully sidelined by colonial powers. Prah notes that we don't talk about 'the indigenous languages' of France or the Czech Republic, but we do so when we speak about African languages (and other native languages of post-colonial societies). This attitude towards native languages is a legacy of colonialism, no doubt. Prah is also involved in a project that seeks to harmonize and standardize groupings of mutually intelligible African dialects as a form of strengthening their foundations and promoting their public use. After all, history has an abundance of examples of how losing a people's language, due to external forces, also means losing their cultures, histories and worldviews (Prah 2009). Most notably English 'linguistic imperialism' currently burdens Africa, says Prah. That is so while we note how the Chinese and others (South Koreans, Japanese, Vietnamese, Turkish, etc.) made visible and strong strides in economic and technological development while (or by) insisting on using their native languages as primary media for communication and education—especially the ones which were colonies for a long time and are now achieving notable national development such as Vietnam and Indonesia.

---

(6)     Interview with Professor Prah by Alicia Mitchel from *eLearning Africa.* May 16, 2013.
        URL:     http://ela-newsportal.com/no-country-can-make-progress-on-the-basis-of-a-borrowed-language/

Yet, a country like Brazil entirely adopted the language of their former colonizer (Portuguese) and yet achieved visible progress with that language (but the Brazilian case has its peculiarities).[7] Countries like India and Singapore chose English to be the primary medium of education and general communication, but also focused on assuring the continuity of their native languages and building capacity for them to continue playing leading roles in communication, trade, media and public policy. An African country, Ethiopia, never lost grip of its native Amheric language, on all public and educational settings, but that did not necessarily render favourable progress compared to its neighbouring African countries that adopted English more rapidly, such as Kenya. Tanzania is another example of an African country that has a good grip on its national Swahili language (which is also an East African *lingua franca*) without it necessarily giving it a visible comparative advantage in terms of economic progress and technological development. It seems that it would be difficult to draw a clear correlation between economic and technological development, on the one hand, and the embrace of native languages on the other hand. However, we should know, such relations are not always demonstrable in blanket statistical linear relations.

Still, however, there is something quite important about learning, sharing and communicating in one's native tongue. There is a deep and wide cognitive infrastructure that connects language to education, communication and collective memory. An Education for All Global Monitoring Report, by UNESCO, 2013/14, confirms that the early years of learning are optimized when using a child's mother tongue (i.e. the language spoken at their home, the first language they begin to express themselves through). Furthermore, while being bilingual and multi-lingual is a historical human skill that often boosts learning abilities and enhances a person's worldview, losing one's native language through years of formal education, and then work life, is not associated with psychological wellbeing and cultural confidence. The possibilities of building strong bilingual and multilingual societies outweigh, in their

---

(7)    An argument about the Brazilian case is that Brazilian contemporary society is similar to Canada, the USA, and other countries where the colonizers settled, reproduced and merged with the native populations (not often voluntarily or peacefully of course). Hence, Brazilians speak Portuguese not mainly as a colonial language, but as an ancestral language of a local majority in power. This is also a case, it is argued, with most Latin American countries.

benefits, those of future monolingual societies in which a borrowed language displaces the history-rich native tongues.

The approach of maintaining and revitalizing native languages, and gradually re-integrating them in economic and technological life, while at the same time assuring fluency in the contemporary global languages of technology and trade, seems to be the most fruitful approach.

> "If language is the main feature of culture, in which all social and human activities are transacted, then we should ensure that our languages become developed and refined instruments capable of yielding developmental benefits in our quest for scientific and technological advancement and…emancipation." (Prah 2009, 7).

## Gender and technology

One of the most crucial social systems is the gendered division of labour. This is particularly crucial in post-colonial societies that are struggling to make a path between imposed westernization and "outdated" pre-colonial native institutions and assumptions. Gendered division of labour is weaved into, and influenced by, many cultural norms and socioeconomic institutions. The cultural norms include perceptions of sexuality, family values, artistic expressions, and codes of behaviour.[8] The socioeconomic institutions include marriage, land tenure, inheritance, leadership and mentorship arrangements, educational structures, and of course physical labour duties.

There are many angles to addressing issues of gender and development, and the literature on this topic is rich,[9] but from the angle of technological development the issue of gender comes mostly on the subject matter of equal representation and agency in shaping technosocial systems that are to be engaged by the whole society. Having only one distinct half of society – typically men – shape and implement technosocial systems that impact all of society is not a sustainable approach. For example, when rural women form the majority of the labour force in agriculture, in a given society, it will not be a sustainable technological change process if men are charged with deciding which technological paths and systems of agriculture are most suitable for the society in question. Important perspectives and expertise are missing from this decision-making and managing process—the perspective of women.

Currently, in a majority of developing societies, we have stark inequality in the representation of women in conceiving, learning, planning

---

(8)    Codes of behavior can be drawn from religion, tradition, or customary/native laws.

(9)    See, for example, Stamp 1990; Fidiel 2006; Lekoko et al. 2012; O'Brien and Williams 2004.

and managing macro and micro systems of technological innovation, application and evaluation. As stakeholders in the technological change process, all involved groups in society have a particular perspective that will not likely be substituted by another group. The gender-based grouping is perhaps the most significant in this aspect because it divides almost the entire society into two distinct groups. Without striving to bring about equal agency to women in all the critical phases of the technological change process, the imbalance will not go away and the results will not be favourable or sustainable.

## The educational divide between technology and policy arenas

Education is a critical social system, more so in a context of development. While this point – the divide between technology and policy arenas – does not surface in most of the literature reviewed for this book, it is a point worth investigating precisely because it does not seem to have received enough attention in the technological development field yet. The gap between policy formulation and executing has one particular problem that needs scrutiny: those who formulate policy are mostly not technology-oriented, and those who execute are mostly not policy-oriented. Much is lost in the translation between the two.

The policy process – from understanding the problem and devising policy solutions to implementing/executing those policies – at any level of work (micro and macro) is faced with the main challenge of connecting policy producers to policy implementers. While the two may have the same goals in mind, 'policy folks' and 'technology folks' operate under different imperatives and with different toolsets. That results in their usual communication channels being characterised by misunderstanding, miscommunication, and sometimes conflicts: "The cultures of the science and policy communities are quite different in terms of timescale, language, academic background and incentive structure, to name a few." (Sheikheldin et al. 2010). Therefore, the task of communicating specific policy concerns to technology people, and making sure the corresponding technology constraints and potentials are communicated back in the right format, is a key and critical task. Morgan (2010) says of this problem in the field of technology and policy interaction that:

> "There is a subset of policy problems in which the technical details really matter – where a failure to consider and address the substance of those details can lead to dumb or silly results…Today many science

and engineering educators are quick to recognize the importance of preparing students with technical backgrounds who can address policy problems in which the technical details matter."

Furthermore, Morgan focuses on how the field of technology-related policy is in need of actors who are conversant in both languages – technology and policy – and who are able to see the elements that weave through both fields. This aligns with the view of some important international actors in the field of technological change, like the UNDP, which concludes that most problems in the technological development arena are essentially about policy, and can be resolved through policy (2001).

In the field of technological change, the need for personnel with combined technology and policy knowledge and skills should be self-evident. Technological development schemes – on the micro and macro levels – should give particular attention to training and fostering change agents, and agencies, that are comfortable with addressing technological problems of policy implications and policy problems of technological implications. Moreover, these agents should be capable of fulfilling the role of effective liaisons who translate knowledge fluently between technology and policy circles. Synchronization and good communication are well known to be essential for good work, and it is no less so for the technology-policy challenges of technological change and development. Training programs of various levels can devise curricula that orient technology folks with policy fundamentals and tools, and policy folks with technology fundamentals and tools.

## Technology and ecological systems

The relationship between technology and ecological systems (or natural environments and their interactions with societies) comes in two dialectic forms:

1. Controlling the environment: where technology is expected to alleviate human difficulties with the elements of nature, to protect and convenience us in the face of climatic threats and challenges, reduce vulnerabilities to disease and food insecurity, and obviate threats from other animals, etc.

2. Maintaining balance in the environment: we realize that resources our livelihoods depend on in the environment are finite, and we ought not to disturb the large, intricate and complex ecological

balance beyond limits where our planet can no longer support our existence.

The description above is crude and over-simplified (details will follow), and serves as a launch pad for understanding our triangular relationship with the environment and technology. At the outset, we want our technology to help us control the environment to our benefit as much as possible. Yet we cannot ignore, whether we like it or not, the critical fact that both our technology and us are products of our environment and part of its ecological equation—and that the overall disturbance of that 'balance' eventually threatens our very existence.

Often when human beings speak of 'saving the planet' they are essentially talking about saving themselves. They are talking about keeping the planet habitable for the *homo sapiens*. And that is okay, more or less. After all we are locked in an anthropocentric perspective of the entire universe and reality, not just planet Earth. Earth is substantially older than us humans, and it can get rid of us in a relatively short period of time, through a series of 'natural disasters' that revitalize the planetary ecosystem in multiple ways, and in the process terminate all or most humans. Earth, as a living existence, will continue while we as humans can become extinct (or significantly reduced in number and capacity).[10] Either way, those who are environmentally conscious are not "selfless" in the way that could be understood from phrases such as 'saving the planet'. Earth, so far, is not threatened with losing its orbit around the sun, or losing its net balance of natural elements due to human activities. Yet environmentalism is necessary and genuine today because we have proven capable of making Earth unfit to our own kind in the name of advancing life for our own kind. As a species we have proven to be intelligent enough to the point of molding many forces of nature to our will, yet we can be unintelligent and short-sighted enough to the point of not collectively realizing that we need to keep a given ecosystem in balance for our own survival.

There is ample empirical evidence today that the actions of humans – cumulated over the decades of the industrial age – are causing Earth to be less and less fit for our life on it, and so far we do not have the option of another planet to migrate to (and in considerable numbers). What is

---

(10)    It is true, however, that the same elements that make this planet habitable to us are the ones that make it habitable to many other earthlings (take most mammals for instance). So there is a good chance they will go extinct like us or significantly reduced in number and possibility of survival.

left, then, is to respond seriously and responsibly to the warnings that have been issued to us by Earth. In the context of developing societies a few more critical factors are added to the equation.

## Low-carbon economies and developing economies

In Africa, climate change is a topic that is of major concern, primarily because scientific consensus has established that Africa is the continent most vulnerable to climate change while being the least able to adapt to its impacts.[11] This means that climate change has, and is expected to have, major impacts on the lives of Africans and their environment. Climate change is also a social problem because responding to it requires social changes.

Climate change has already started to affect the continent. According to the Intergovernmental Panel on Climate Change (2007): "In the [Sudanic] Sahelian region of Africa, warmer and drier conditions have led to a reduced length of growing season with detrimental effects on crops. In southern Africa, longer dry seasons and more uncertain rainfall are prompting adaptation measures." For obvious reasons, the economic resources of African countries are directed towards a list of priorities other than climate change. For example, while growing forests mitigate climate change, things are not that simple on the ground. Livelihood demands, especially for rural communities, make over half of the African population, to date, dependant on forest consumption for fuel-wood. There is also increasing demand for energy from the cheapest and most polluting resources to meet the basic needs of economic development that are yet to be fulfilled. Such conditions indicate why adaptation to climate change in Africa is one of the toughest environmental and developmental challenges to the present and near future generations.

Carbon dioxide emitted during human consumption of fossil fuel energy, as an unintended consequence, is considered by far to be the single biggest contributor to climate change. This consequently calls for the need for strategies to reduce carbon emissions on a local and global scale. A low-carbon economy, or a decarbonised economy, is understood as one where the supply chain of products and services is attentive to its carbon footprint and effectively integrates changes that lower that footprint to significant degrees.[12]

---

(11)    Parry et al. 2005; Kandji et al. 2006.
(12)    See Delay, T (2007). "Low-carbon Economy: What are the opportunities?" The EIC guide to the UK environmental industry. Retrieved October 14, 2013 from: http://www. eic-guide.co.uk/docs/lcarbon.pdf, and Greenpeace (2010). "Decarbonised Economy:

In scenario forecasts and computer modelling that help foresee the impacts of different approaches of adaptation to climate change, technological means of adaptation are generally not included as a factor (Devereux 2004). This means that the climate change debate often under-emphasizes possibilities of future technological advances, changes and breakthroughs as solutions for climate change. While it is understandable, to a degree, that one cannot base scenario forecasts on the big assumption that technological innovations will appear and become game-changers, this always leaves us with the possibility that technological adaptation to climate change is an area that should continue to be explored.

Investing in technological R&D, promotion, and implementation can prove to be highly critical for shifting climate change mitigation strategies towards more optimistic scenarios. For example, France is hailed by many as an energy model of the future due to the fact that 75% of its electricity is generated from nuclear energy stations, which makes France the leading low-carbon economy among the industrialized countries and arguably in the world. In addition, the Canadian province of Ontario generates 50% of its electricity through nuclear power stations. The Ontario Power Generation Corporation states that nuclear energy "has two major benefits - low operating costs and virtually none of the emissions that lead to smog, acid rain or global warming."[13] Other nations may have to take bolder policy decisions and shift to nuclear energy, especially seeing as the popular perceptions about the dangers of nuclear energy do not match the empirical data and the majority voice of technical experts (OECD 2010). And while solar, wind, tide and geothermal energy sources are not yet strong, affordable, or efficient enough to replace fossil fuels, there is certainly more room for improving the quality and quantity of these renewable resources. Germany, for instance, has been able to produce 18 billion kilowatt-hours from solar photovoltaic energy in 2011, and is planning for a target of 35% of its power generation from renewable energy sources by 2020. Some sources are even hopeful that 100% of Germany's power – with the right technology and policy combination – can be generated from renewable

Opportunities and responsibilities of the ICT sector in a changing climate." Retrieved October 14, 2013 from: http://www.greenpeace.org/india/en/news/Decarbonised-Economy1/

(13)   Ontario Power Generation Inc. – OPG (2000-2013). "Nuclear Power". Retrieved October 15, 2013 from: http://www.opg.com/power/nuclear/

energy sources by the year 2050.[14] Other parts of the world which are endowed with more solar and/or wind energy exposure than Germany – such as most parts of Africa – can invest more on that track. The main recipe that is required for these types of responses to climate change to work, has both technology and policy as critical ingredients.[15]

Positive aspects of the low-carbon economy concept are that it addresses carbon dioxide directly as a main target, and that the concept seems to support continuous investment in technological R&D for reducing carbon emissions. Some problems exist, however, with the concept and its indicators. One problem is that it obscures the reality of challenges to sustainable and equitable economic development around the world. The terminology of low-carbon economy, because of over-generalization, does not say anything of value about the reality that within each national or regional economy there are unequal consumers; hence unequal polluters. In addition to that, indicators such as carbon emission by unit of GDP further propagate the process of ignoring the existence of unequal polluters. While some countries can be considered low-carbon economies due to their conscious investment in energy production without greenhouse gas emissions, other countries are low-carbon economies due to the fact that they are still struggling economically to provide sufficient energy to the majority of their citizens. This means that, from an equitable economic point of view, you would expect the former countries to decrease their carbon emissions even more, while you should expect the opposite from the latter countries. Even within big-polluter countries that are expected to lower their carbon emissions that responsibility does not rest evenly upon all the citizens, since they are not equal in their contribution to the national level of greenhouse gas emissions. Another problem is accounting for the amount of pollution by transnational corporations that are based in rich industrialized countries but carry out their heavy-pollution activities elsewhere around the world. The concept of the low-carbon economy can sometimes obscure these complex details.

Perhaps a number of alternative policy perspectives can retain the positive aspects and resolve the negative aspects of the concept of low-carbon economy. Thinking about the problem of climate change

---

(14)     Paul Hockenos (2012). "Germany's Grid and the Market: 100 Percent Renewable by 2050?".*Renewable Energy World.com*. Retrieved October 15, 2013 from: http://www. renewableenergyworld.com/rea/blog/post/2012/11/ppriorities-germanys-grid-and-the-market

(15)     See OECD 2010b.

in developing societies from the angle of low-carbon economy can be useful as it addresses environmental, economic and technological matters of development all at the same time, and in a realistic fashion.

The current global ecological challenges pose difficult questions, such as how can developing societies in the world today achieve prosperity for the majority of its inhabitants who are currently living in unfavourable conditions, without taking the same path of modernization that was taken by the industrialized regions of the world—i.e. without human exploitation and environmental pollution? And with all the many pressing problems to worry about, is there room for developing countries and communities to prioritize matters of responding and adapting to climate change? How would that rank with respect to "other" pressing priorities of education, health, economic self-reliance, and peace and security? How would and should resources be organized to address these overwhelming matters all at once? What changes could be made to the concept and framework of the low-carbon economy to suit the contexts of developing regions? And what role would technological solutions play in these changes? These are all questions that have technology-institutional dynamics at their core.

### Technology, nature, and commodification

The concept of fictitious commodities, introduced by Karl Polanyi, and explained earlier in chapter three, presents a useful analytical tool to look at the relation between technology and nature in modern societies.[16] The ways in which market economies impact modern societies appear in how they create a divorce between economic institutions and other social ones. This divorce sees economic institutions rendered within the purview of 'rational choice theory' while social institutions adhere to wider values and cultural patterns. Polanyi calls this divorce phenomenon 'disembeddedness' (i.e. economic institutions are no longer embedded in social values and structures). The way the market economy creates this state of disembeddedness is mainly through the creation of fictitious commodities, including land, labour and money as was detailed.

This Polanyian narrative gives insight for the analysis of socioeconomic phenomena that include technological elements. The notion that land (natural resources) and labour (human activity) are not commodities provides a good 'environmentalist' lens to economic

---

(16)    See Polanyi 1944 and 1968; Hopkins 1957; Harvey et al. 2007; Harriss 2003.

and technical discussion, from which the topics of environmental degradation and climate change, due to human industrial and consumerist activities, can be examined. We observe that the two dominant factors in making and utilizing technology are land (natural resources) and labour. Both are fictitious commodities. It was predicted by Polanyi in 1944 that this tendency of the market economy to commodify natural resources and human labour would eventually lead to problematic results. He said that, "such an institution could not exist for any length of time without annihilating the human and natural substance of society." Our modern economic cycles, which integrate nature and technologies, are proving to be in need for serious revision.

## Technologizing a better way

Yet technology can very well be part of the solution notwithstanding that it is today part of the problem. Technological innovations and promotions that are more sustainable and more ecologically sound continue to emerge. In addition, many traditional technologies that had been neglected for years are now being reconsidered, by local communities and by research institutions, for their robustness and environmental adaptability. Rather than totally bypassing these local technologies they can be revisited, re-modified and re-contextualized to serve communities better.

In the Sudano-Sahelian region of Africa there are examples of organized efforts that work with communities to coordinate, organize and improve a number of traditional water harvesting systems.[17] From northern Ethiopia, Mintesinot et al. (2004) report on case studies of water harvesting projects led by the Ethiopian authorities, such as micro-dams and diversion weirs. Other countries in the Sahel have similar experiences with water management systems.[18] One of the traditional water management systems in Mali, the Zai system, has been credited with noticeable success in combating desertification and increasing agricultural yields.[19]

Another example comes from the Middle East, particularly Oman. The administration of traditional water systems in Omani villages

---

(17)  Such as *hafirs*, contour schemes and terracing in Northwest Sudan, combined with non-traditional check dams along runoff streams, to collect and preserve water from the flood season to use in the dry season (Fidiel 2005; Siddig 2004; Practical Action 2002).

(18)  See Taber 1995; Barry et al. 2008; Kabo-bah et al. 2008.

(19)  The Zai system is comprised of water harvesting techniques for agriculture, and "meets the criteria for three conservation practices: soil conservation, water conservation and erosion protection." (International Federation of Agricultural Producers 2005, 5).

(*aflaj* systems) follows a community-based, non-market pattern. As Al-Ghafri (2008) explains, "Falaj (singular of aflaj) is a canal system constructed above or underground to collect underground water, water from natural springs, or water from the base flow of wadis. Aflaj provide water to farmers for domestic and/or agricultural use". This system is still prevalent today in the rural regions of Oman and is quite efficient. The Omani state authorities have been wise enough to embrace and refine the *aflaj* systems, lending a modern technical hand when needed, instead of neglecting them for 'old hacks'.

Modern technology, meanwhile, such as renewable energy and energy-efficiency technologies have been improving in quality and increasing in quantity. Industrial technologies and systems continue seeking to implement new measures and innovations to reduce their industrial waste. One of the main sources of pollution on Earth is the industrial sector, especially in industrialized countries. Products and production methods beget a big portion of hazardous waste to the environment through factories and consumers around the world. Besides air and water pollution by factories as a direct residue of industrial processes as well as the disposal of un-reused, un-recycled and un-recyclable finished products after consumption, industries produce huge amounts of waste daily, in the form of unwanted raw or even fabricated material that did not make it to be finished consumer product. This type of waste is completely undesirable, for both the environment and the industries, because besides polluting the environment and depleting energy, waste generates no revenue for industries. A question must then arise: why is this continuing to happen when it is clearly neither doing any good for the economy nor for the environment?

Many industries, for instance, do not yet have clear ways of eliminating industrial waste, or do generally know of ways but do not use them. They perceive industrial waste as a necessary by-product of their production processes. Industrial waste comes mainly from deficiency in the production systems, in machine inaccuracy, and human error, which are all results of the imperfection of human knowledge and implementation. It is true that production systems have evolved greatly over time, and especially in recent decades, resulting in greater capability in mass production, more accuracy and quality assurance, and containment of human and machine errors, but some may say that that does not mean that we can ultimately eliminate, or significantly reduce industrial waste yet. It is a valid argument, one might add; compromising economic growth for the sake of the environment has never been intuitive for human societies after all. The fallacy appears, however, in another aspect, which can be summarized in the question: do we really have no existing tools and systems that can significantly reduce industrial waste, or even eliminate it at some levels?

Industrial waste and cost reduction systems, such the Toyota Production System, Lean Manufacturing, Six Sigma certification, and other quality assurance and control procedures are already in use in many industries today. These industrial systems have already provided considerable efficiency in practical reduction of waste and energy for firms that adopted them, and they continue to improve. These methods are applicable, and they need more initiatives and improvements. Industrial places and firms in developing societies could adopt them as fitting to their conditions. Some public regulation can and should be used here. Experts in local, national and regional governing bodies with the proper familiarity with both technology and policy can introduce regulations that push industries towards adopting waste-reducing behaviour. Mass communicating the economic benefits of implementing these industrial systems has its rewards.

It is certain that reducing industrial waste is not enough to solve the global environmental problem in general. On the other hand we also need to change our consumption behaviour to be eco-friendly, and reducing industrial waste does not address this problem. Yet, all things considered, reducing industrial waste is an integral part of a coherent and comprehensive approach. Addressing the issue of overproduction is a continuing one, whether we have more eco-friendly products and processes or not, because overproduction is always undesirable, for the economy and for the environment.

Finally, technologizing better ecological and social systems is possible, on various fronts, because of human ingenuity. I would not mention examples from all the possible technology sectors—e.g. modern transportation, biomimicry, industrial ecology, etc. Technology, as we have learned from experience, is capable of doing wonders, as it has many times before, when ushered by strong, reasonable and sustainable foundations. To echo David Suzuki:

> "The human brain now holds the key to our future. We have to recall the image of the planet from outer space: a single entity in which air, water, and continents are interconnected. That is our home."

# Chapter Five

# Technology and Justice

*"We can try to cut ourselves from our fellows on the basis of the education we have had; we can try to carve out for ourselves an unfair share of the wealth of the society. But the cost to us, as well as to our fellow citizens, will be very high. It will be high not only in terms of satisfactions forgone, but also in terms of our own security and well-being."*

–Mwalimu J.K. Nyerere

Why Justice? Because justice is central to the topic of this book. The goal of development, particularly sustainable development, is getting human groups to reach the better life and livelihood conditions that they deserve. It is based on a value judgement on human life and what its conditions should be in terms of assurance of needs satisfaction, reasonable levels of comfort and room for expression, and for aspirations. These are not matters of pure material progress, but also of management and distribution of resources. Justice is central because it focuses on how each human individual and group gets a fair share in respect of meeting these aspirations. If those conditions are fulfilled for some individuals and groups at the expense of others then there is no sustainable development; not the development this book is talking about.

This study, particularly in the context of development and sustainable development, approaches justice not as an ideal state to which we compare other conditions and seek to make them measure up. This approach is useful sometimes in philosophical undertakings that seek to build theoretical foundations and/or a consensus on the idea of justice. However, in examining real life conditions, of history and the present, this approach has the shortcoming that there is no place on earth, or a society in all history, that we could speak of as having realized justice. Justice could not be agreed to have been realized anywhere in the world in any period of time. Similarly, there are degrees of differences, and we could argue that some societies are 'more just' than others, but that argument would be highly contextualized and it would be arduous work, if not useless, to replicate it for other various contexts. In the discussion of specific conditions or specific aspects of justice in developing societies, a fruitful approach will be to steer a considerable distance from these theoretical complexities.

Instead, this book approaches the pursuit of justice as a pursuit of exposing and combating situations and occurrences of injustice. It may sound circular at first glance, but there is a technical difference between identifying justice, as an ideal state, and identifying injustice as real occurrences. In any given context where we know the main topic and the elements of our focus, we can detect injustice, and agree that this is so, more effectively than measuring things up to an ideal definition of justice that we have to agree upon first. Conceptually, it is true that injustice is simply the negation of justice, but practically we as humans are better equipped with a sense that 'something is wrong' than a sense that 'this is perfectly right'. Consider the analogy of an empty room—ideally, the room may appear to be empty but it could be argued that it is not really 'empty' and that argument would have merit (i.e. there might be air, microorganisms, material for walls and floor, etc.). However, if we define 'emptiness' as an absence of furniture, then this would be more effective to establish agreement that the room is empty.

In the context of developing societies, with respect to matters of development and technology, we can all recognize injustice when we see it. When there is a clear case of disproportionate harm, dispossession and alienation, to a group of people on aspects related to technological development, we can see it and see why it is so. We could argue a more constructive case around it and address its origins, outcomes and possible removal. In the rest of this chapter we will demonstrate a number of such cases, selected carefully for relevance to the theme of the book.

# On progress, modernity and westernization

There is a particular dilemma that has been faced, and continues to be faced today, on multiple levels, by natives from developing societies who preach and promote modernization in their societies. Of course there are many natives who do that, and a large number of them have received relatively high education with certifications from Western-style institutions (local or foreign). Many of the leaders of the anti-colonial liberation movements around the third world belonged to this category, and nowadays many native promoters of change in their countries belong to that category as well.

Granted, this group of natives has never been monolithic. Their ideologies, aspirations and tools of social engagement are diverse. Here, however, I would like to focus on the individuals from this group who were/are engaged in legitimate processes of decolonization (as defined earlier in this book) and who also promoted/promote selective learning from the relatively positive aspects of the colonial and western legacy in order to forge modern societies in their homelands—i.e. those who are simultaneously involved in promoting decolonization and modernization.

Often those persons, as individuals and as active minority groups (the intelligentsia, perhaps), have been faced with attempts of social ostracism and isolation from the rest of the population for being 'westernized'. Their propositions were/are often discredited for being un-innovatively borrowed from other societies and thus carrying an intrinsic inferiority for the receiving societies. Their attempts and approaches to transform certain cultural assumptions and practices – as judged to be hindering the progress of their nations – were often met by criticism from wider native conservative forces, who accused them a lack of respect and appreciation for their own cultural heritage and native history. These individuals and groups have also been reminded more than others that the societies of the former colonizers – i.e. western countries – are also full with problems and failures, and that they are not perfect. Later on, post-modernist critiques from western writers and observers alsotook turns at criticizing this same group of natives of post-colonial societies. Post-modernist narratives of the 'now-understood' traditional systems and values of the pre-colonial societies speak of how they were right all along - the traditional pre-colonial systems - and that the western-educated native elites knew little about their own native systems and histories because their minds were colonized.

It is a difficult task to vindicate some of these individuals – the native intellectual elites. The accusations themselves hold some truth to a degree, but they apply to various individuals and intellectual groups differently. Some of these native elites have been guilty of those accusations to a considerable degree, others less, and others do not deserve such accusations at all if we look at the big picture. Yet, if a blanket position on the accusations is to be taken, then it is that the accusations should be rejected; not because they are fundamentally wrong, but because their foundation is questionable. The logic of the vindication of the progressive native elite comes from understanding history as a network of events and phenomena in which new conditions are born from previous conditions, not from vacuums. The vindication case is presented in the following argument (presented in numbered points):

1.      We know and fully understand that colonization did more than take over lands and subjugate their peoples. It also heavily interrupted and replaced the local social evolutionary processes.[1] The era of European colonization was not similar to previous versions of occupation and conquering in human history. In the previous versions, occupation would often entail massive subjugation and transformation in power structures. People begin to answer to new authorities, and their labour becomes summoned differently. Also the fruit of their labour may benefit new masters, and their markets may witness some shifts. However no massive technological replacement takes place with regards to how they produce their food, build their shelters, and exploit other resources of the land. Gradual technological changes may happen, but not massive shifts in quality and quantity.[2] Also their socioeconomic and cultural institutions that do not conflict with the new conqueror's immediate terms of rule are not threatened with being

---

(1)     For Africa, it was both colonization and the slave trade. The trans-Atlantic slave trade was exceptional in history in terms of its long duration (spanning over three centuries), the massive number of people victimized (estimated at around 20 million men, women and children, the size of multiple entire countries especially at that time), and the legal and intellectual legitimacy it was given through biased readings of Judeo-Christian scriptures and some pseudoscience claims.

(2)     "Given the fact that African colonies were organized and viewed as sources of raw materials and potential markets for manufactured products, colonies were accordingly structured to ensure that they did not develop a sense of national identity. Forms guaranteeing dependency were initiated. Indigenous technology was brushed aside while there was reluctance to transfer and adapt appropriate exogenous technology to the social, cultural, economic and environmental structure and needs of the people." (Forje 1998, 11).

uprooted and left to die. Their languages will not be massively replaced in all aspects of their lives. That was the general picture of previous histories of occupations. European colonization was different. It was more extreme and pervasive in its effects. When Europeans conquered the societies in Latin America, Africa and Asia, there were functioning civilizations and local social evolutions taking place. These evolutions and civilizations would have continued along trajectories that would see further unlocking of their own productive forces and contradictions, leading to their own improvements or transformations or replacements, as is the historical story of social progress. Those trajectories were virtually completely halted with the arrival of modern colonization.[3]

The particular difference of European colonization was not necessarily due to Europeans themselves, however, but largely due to the fact that those were the times when astronomical differences in technological capabilities took place; i.e. contemporary to the discovery of the fatal power of guns as well as the emergence of the industrial revolution in Europe. It is evident from history that the industrial revolution was a result of cumulative knowledge acquisition by humans from all over the world, but the decisive moments of transformation took place in 18th century Europe. That historical happening – irrespectively of the factors that aligned together to make it happen in a particular region of the world at a particular period of time – gave Europe an undeniable advantage over the rest of human societies from that time up to this day (although that advantage has been steadily fading in the recent decades).

2.    Furthermore, colonialism was conscious in taking measures to prevent colonized societies from picking up the colonizer's technological knowledge through a natural historical process of interaction and gradual learning. It is documented, for example, that Governor of Tanganyika, in 1935, Harold McMichael, quoting a dispatch from the Secretary of State for the Colonies

---

(3)    Diop (1988) recounts the many advancements in technology and social organization that were taking place in various African societies prior to colonization. We also know about the African civilizations of Meroë, Axum, Zimbabwe, Songhai, Dahomey, Swahili city states, Egypt and others, that witnessed endogenous technological and institutional transformations in metallurgy, architecture, agriculture and trade logistics for centuries before the latest corrosive contact with expansionist Europeans. Such transformations are indeed seasonal byproducts of humans interacting in social contexts anywhere around the world.

of the British Empire, said, "It was undesirable to accelerate the industrialization of East Africa which must, for many years to come, remain a country of primary produce." For economic and political reasons that are not difficult to guess, it was an intentional policy to keep the colonies technologically behind the metropole of the colonial empire. In 1729, English merchant Joshua Gee, in a then best-selling book ("The Trade and Navigation of Great Britain Considered"), expressed the same policy orientation in more detail. "We ought always to keep a watchful eye over our colonies, to restrain them from setting up any of the manufactures which are carried on in Great Britain", he wrote, "and any such attempts should be crushed in the beginning, for if they are suffered to grow up to maturity it will be difficult to suppress them". He even clearly advocated that "all Negroes shall be prohibited from weaving either Linen or Woollen, or spinning or combing of Wool, or working at any Manufacture of Iron, further than making it into Pig or Bar iron: That they be also prohibited from manufacturing of Hats, Stockings, or Leather of any Kind... Indeed, if they set up Manufactures, and the Government afterwards shall be under a Necessity of stopping their Progress, we must not expect that it will be done with the same Ease that now it may." His logic was to preserve "the advantages to Great Britain from keeping the colonies dependent on her for their essential supplies."

This was a universal policy of European colonization, even when not overtly expressed as in the case of the British Empire. In his seminal book *How Europe Underdeveloped Africa* (1972) Walter Rodney provides one of the most articulate and enduring arguments, with detailed historical evidence and analysis, for the following:

- That European colonization of Africa contributed directly to the chronic material underdevelopment of African peoples in their continent. Many of the development problems that Africans still struggle with to this day can be objectively traced to origins in colonial times and colonial actions. Africa was undergoing processes of historical development, in systems of governance and production, before Europe's negative intervention halted those processes.

- That Europe directly benefited, in clear and large measures, from exploiting African labour and resources, to fuel its own leap to unprecedented wealth, overall prosperity and industrialization in Europe and European settler colonies. Africa clearly contributed to Europe's capitalist development.

- That throughout that process, Europeans committed systemic, unspeakable crimes against humanity in Africa, even by the standards of those times. Evidence of the moral failure and hypocrisy of colonialism could not be more vivid than it was in Africa.

- That the general transformation in Africa, during colonial times, to modern infrastructure and new forms of administration and public services, was mainly for the purpose of serving European colonial interests, not the colonial subjects. Any byproduct of relative material benefits, that was extended to a relatively limited number of native Africans, does not stand in objective scale compared to the opposite. When it was not directly beneficial and did not have direct enjoyment and productivity impacts for Europeans, the efforts that were put into 'modernizing' the African landscape was kept to minimal and symbolic measures.

- That many of the positive trends and events that happened during colonial times and postcolonial times can be largely credited to the direct and indirect impacts of resistance to colonialism, not to colonialism itself. Some good things happened during those times despite colonialism, not because of it. When Africans snatched what they could from the knowledge, arsenal and culture of the colonizers, and used that for their own benefit, they demonstrated that they were able to begin a process of reclaiming their own development under terrible conditions.

- That whatever the current challenges, successes, shortcomings and failures that Africans are going through today (as many post-colonial societies are), in the grand process of re-developing and reclaiming their place in modern times, any talk about favouring aspects of colonial times over current times are at least misinformed and misguided, and at most a historical and immoral.

- Rodney's narrative is not without challengers, of course, but his argument, and supportive evidence, endured over time. It does not explain the whole picture (particularly the present picture, decades after political independence was secured) but it sheds light on important aspects going forward.

3.   Points 1 and 2, show how a severe evolutionary disconnect with their pre-colonial history happened in native societies. As Amilcar Cabral (1966) expressed, societies that experienced severe and prolonged colonization were essentially taken out of history. Their history became only an extension of the colonizer's history for a period of time. Returning to their own history again, through the process of decolonization, is no easy affair. In their post-colonial times these societies only have glimpses of what their own social dynamics were like. Post-colonial societies have to deal with the very peculiar task of gathering, dusting off and resuscitating whatever left of their pre-colonial identity while having to do that under new conditions of existing structures of governance and politics (i.e. the European modern state and modern international politics), structures of production (i.e. European technologies and markets), and structures of communication (i.e. European languages and procedures). It is a task lacking in any particular envisioned ideal, and in the brightest versions of possibility will end up creating something very different from both pre-colonial societies and colonial times—something that never existed before.

4.   We do not know how pre-colonial societies would have evolved without colonial interruption. We do not know the path their evolution would have trodden. We do not know how that story would have unfolded, because it was not allowed to unfold. No one – neither the natives, nor the former colonizers nor others – can offer a version of post-colonial society as if colonization never happened. What we can be sure about, however, as a rule of thumb, is that pre-colonial societies would still have evolved without colonial interruption. They would have become by now something quite different from what they were back then. We cannot assume that, had there not been any colonial interruption, these societies would have stayed the same. That would be a most unhistorical conclusion. Human societies do not stay static, for dynamism and change are in their nature. In addition, external influences continue to happen to societies without colonization—through trade and other forms of interaction. It is almost certain that pre-colonial societies would have changed in many ways without colonization (just not in similar ways as under colonization). This observation informs our dealing with pre-colonial history: it would have been 'history' by now either way. It would not have been our 'present'

either way. We do not owe colonization that default trajectory towards progress/complexity.

5.  At the existential level, we are the products of our cumulate story. Not only are our conditions a result of that, but even our imaginations and aspirations. When post-colonial societies started to dream about what their independent nations will be like, they had no choice but to vaguely imagine some sort of mix between the pieces left in their collective memories of their pre-colonial life and some of the colonial legacies that were considered good or inevitable. There was no possibility of imagining something other than some mixture of two histories that are themselves worlds apart and each with its own bitter taste in the mouth—the bitter taste from pre-colonial heritage being the legacy of absolute and everlasting defeat, and the bitter taste of colonial times as times of massive oppression, dispassion and alienation. Even more so, that imagined mixture could not have been escaped while the former colonizer is still very present in the post-colonial life— in diplomatic relations, trade relations, industrial relations, and under a global community governed by the same institutional characteristics of the former colonizer.

6.  All the above considered, it would have been inconceivable to imagine that leaders of post-colonial societies would pursue futures for their new nations that did not include heavy elements of 'western' modernization. What else was there for them to imagine? While their proposed national modernity projects included an inherent critique and rejection of many aspects of European modernity, that critique was there primarily because their modernity could only be with some conscious deviation from European modernity, not an absolute negation of it.[4]

---

(4)  It is worth emphasizing that post-colonial modernization thinking and planning, as promoted by the majority of leaders in developing countries – both intellectual and popular leaders – is inherently an active critique of colonial and European version of modernization. Anti-colonial liberation movements left a legacy of viewing European modernity with a critical eye—learning from it but refusing, on principle, to emulate it. That is well documented and is no surprise, since liberation was essentially against imposing a European way of seeing the world and interacting with it; a way that centralizes Europe and its people as masters and leaders in everything. That foundation of principled critique of European modernity may not have always manifested in clear action after political independence, but it remained there as a foundation that all agreed upon in theory. It was, and still is, never challenged in principle, which is what makes the discourse of modernization in post-colonial societies inherently critical of the colonial-European approach to modernization.

They did not envision that their new nations would be the anti-thesis of European modernity, but a new synthesis of their own determination between European modernity, pre-colonial heritage, and something totally new and unique that is their own colonial experience and the reflections that came from it. In any case, just as the industrial revolution and its outcome was essentially a result of cumulative history of human invention and innovation worldwide that culminated in Europe at the time, there is no doubt that the positive aspects of modernization are also the rightful claim of all of humanity.

7.     Hence, we can see that it was not necessarily 'westernization' that guided the visions, decisions and actions of native elites, especially anti-colonial liberation leaders, and not necessarily westernization that is guiding their successors nowadays. It was simply history that was and is guiding them. As one of them once expressed:

> "The process of development in the advanced capitalist countries took a historical route which we will not follow...The industrialization of Europe was not an easy affair. It involved colonialism, it had involved exploitation including that of child labour...In Africa, we cannot go out to colonize other places. This is out of the question; there is nobody else to colonize."[5]

If these leaders make mistakes, here and there (as indeed they have and probably still will), with that particular synthesis vision in mind, it is not necessarily because they were overly influenced by western ways. They could have just made mistakes of judgement, of strategy, of analysis, etc. – the kinds of mistakes that must take place when you are taking on a task as large and comprehensive as forging new nations from old complexes. There are many factors to be considered, and many ideas to be reassessed and re-thought, but taking refuge in blanket critiques, such as that 'they were too westernized', is not justified, whether these critiques are offered by other natives or by western scholars (so-called orientalists or post-modernists).

8.     There are many benefits that modern technology and systems provide to peoples' lives and livelihoods that are inherently matters of universal human preference, not Western influence. For example, human societies always strived to see better and have

---

(5)    John Garang de Mabior. Excerpt from a seminar he held on June 9th, 1989, Brookings Institution, Washington DC. See de Mabior 1992.

more illumination at night. They also always sought to find faster and safer ways to move between places and move their belongings with them. They always sought to exchange some of their crafts with the crafts made by others, because there were mutual interests in the products and commodities that others may have learned to provide in better ways. They always sought better health and immunity to various diseases. They always sought to make food and water more accessible, and ensure further safety from natural enemies and the elements. They always wanted more comfortable and functional clothing and tools. They always wanted to be able to communicate with each other across distances more conveniently and effectively. Human societies made many advances in these desirable aspects over long histories, and they made them through either independent discoveries or learning from each other under various circumstances of interaction.[6] It is no surprise therefore that even before the industrial revolution there were shared technologies throughout the world in agricultural and irrigation processes, masonry and architecture, travel and carriage, medicine, weaponry, writing and printing, etc. Modern technology added significantly to making these universal desirables attainable. For people outside the West, wanting these fruits of modernization means wanting these things, not necessarily the whole western package that comes with it. Yet, being a social product, modern technology comes with certain conditions and arrangements (institutions) that may have crystalized in the West; some people accept these institutions or seek to adopt them as much as is needed in order to benefit from modern technology. To want that, and to want that for one's people, are not necessarily signs of being westernized. Ironically, most of those who criticize 'westernization' of native populations through modernization do not mind enjoying the benefits of modernization themselves.

As John Garang (1981)[7] once said, "Indeed the people in the rural areas [of developing societies] appear to be more interested in change than some of their educated brethren who appear to be

---

(6)    Some of these circumstances of interaction were not the preferable of all sides, such as war and conquest. Others were, such as trade and alliances.

(7)    John Garang de Mabior (1945 – 2005) was the leader and co-founder of the Sudan People's Liberation Army/Movement. The quote is from his doctoral dissertation in agricultural economics, Iowa State University, which was on rural development in Sudan and South Sudan. Page 6.

mission-bound to protect [the traditional] "way of life" which they themselves appear to have rejected."

9.   With the understanding above in mind, the critical perspective this book takes on some of the failed industrialization schemes in Africa – part of the technological transformation schemes for modern African countries – is a critique of vision and practice without levelling the often unfair and unfounded accusation of 'too much westernization'. After all, it is still difficult to imagine a future for a more developed post-colonial Africa, or South-Southeast Asia or Central-South America without imagining a comprehensive presence of technologies and ideas that are, for the sake of convenience, it has been shown, described as western. Indeed many of the transformations that happened in modern history – which happened to take place in a time of European global dominance – were simply transformations in universal human and social elements and can be claimed by all those who lived through them and engaged them in any part of the world. Yet, with that in mind, new and different societies from the western model, informed and inspired by both pre-colonial history and post-colonial experiences, are surely attainable and legitimate. Indeed, there is room now to foresee and conceive a future where the West will itself be transformed by ideas, perspectives and even technologies that emanate from other parts of the world.

## The Individual vs. Society

Moving forward with the process of scanning and clarifying ideas on modernity and progress in developing societies, we find it important to touch on the major topic of the dialectical relationship between the individual and society.

The argument that 'society is more important than the individual' is an old one. It is also claimed that most pre-colonial societies adhered to it as a founding norm. It is argued that the modern trends of securing and enhancing individual freedoms are generally alien to native non-western societies. Yet we know that very recent and different western regimes and empires adhered to the principle of 'society over the individual'—e.g. the Soviet Union, Fascist regimes, war-time governments of erstwhile democratic countries, and of course colonial governments.

The argument for the dominance of societal interests over individual interests starts by tracing a clear dichotomy of 'society vs. individuals'. Then it argues for the favouring of society over the individual, using certain cultural sentiments such as family structures and specific religious/cultural beliefs, then elevating them to use within socioeconomic institutions such is most apparent in gendered divisions of labour, tribalism and land tenure and division of power, and so on. In the realm of political philosophy the communitarian versus liberalism debate virtually maps this reality. We will, however, not discuss here the historical philosophical records, from across the world, which discuss this old dichotomy of society vs. the individual, for that will be too much of a deviation from the topic of this study. The point to be taken, however, is that it is safe to state here that it is an old discussion, which means that it was never taken for granted throughout history in all pre-colonial societies. It has always been a contested idea, and more often the contestation was not simply philosophical or abstract but rather about deciding how much room individual choice and privilege should be acknowledged and protected by social codes and social authority. There were no absolute decisions taken throughout social history, but rather contextual decisions. Taking that narrative into consideration we can today re-open and discuss the matter through new outlooks within post-colonial societies, due to many changes in the contexts, and not be accused (again) of simply mimicking the West.

The age-old discussion starts from the premise that individual rights and freedoms need to be suppressed – by force if needed – for the sake of the security and continuity of society, because the continuity of society at large surpasses the importance of one individual or a handful of individuals. The advantage of the big group takes precedence over the convenience of individuals. It is a matter of survival and welfare of thousands compared to tens. It is logical and quite intuitive, the argument claims. Limited privileges can be offered to individuals in society as conditions permit, but these are subject to society's determination – which means they can also be recalled; i.e. 'society giveth and society taketh away'.

The response to this blanket argument comes in steps. We first observe that the norms of society are always dynamic, never static, and that they are in any case always carried on by individuals, not by groups. We must always remember that social groups do not have a collective heart and mind. Minds and hearts belong to individuals, always. This means that any prevailing opinion and mindset in society is essentially reflective of the group of individuals who make their views dominant in their society (whether by consensus or by force).

Yet the ultimate purpose of society is not to reproduce itself. It is rather to foster individuals. Individuals engage in 'social contracts' for their own interests, not for the interest of a vague entity – which has no soul and ambitions of its own – called society. Therefore, it follows that if society fails to foster individuals then it is a failed society. This is not to say that society can never have interests other than those of individuals, because there are collective/shared interests and private/personal ones. For example, while everyone in society would agree that they want security, physical infrastructure, food and public services for themselves, these qualities can only be achieved by a coherent social system that could administer a big network of resources and labour. However, some individuals think they can have or enjoy these 'goods' and at the same time be absolved of the responsibility of contributing their fair share to creating and sustaining such goods. These individuals still want to 'keep their own wealth' to themselves and have no boundaries on their freedoms (even if they may result in violating other people's freedoms). These individuals simply want to be well-accommodated by the group without having to reciprocate the favour. These are unrealistic and irresponsible individuals, and irresponsibility does not go well with freedom, in any environment. In such cases, we have an immature understanding of individual freedom. Those who want to enjoy the benefits of living in society without sharing in the efforts to build it should try living outside of it! It is not freedom to be parasitically dependent on the group when you can be a positive contributor (in whichever capacity available to you) unless there are genuine objective conditions that constrain your ability. Important measures of social justice need to be established in order for individuals to realize their true freedoms and potentials.

Individuality – our personal treasure – expresses itself in social contexts. Without society, a manifestation of individuality is virtually meaningless. Here lies the difference between 'chaotic autonomy' and 'individual freedom'. A free individual recognizes that social good enhances the potential of their own individuality.

However, there is, realistically speaking, a balance that can be pursued, and relatively achieved, between individual freedoms and societal interests. This balance is called 'the constitutional framework' in modern legal parlance. It is not yet perfected, but it is far more productive to engage in defining and implementing this framework, or getting the balance right, than to take the extreme polarized position of 'individuals vs. society.' That is in fact one of the chief tasks of modernization for developing societies (An-Na'im 1995). Naturally, since modern societies are weaved together through technological products and systems, the task of fostering and sustaining technological development is part and parcel of the task of modernization.

## Global relations of contemporary technology

The industrial revolution, which began in 18th century Europe as small but qualitative breakthroughs in manufacturing techniques and machinery and then grew to full-scale transformation of industrial processes, marks the identity of our modern age. It is not that technology was not an important feature in most human societies before then. It has always been. But with the industrial revolution it became the defining feature, the most important, and the one that makes all the difference. Before the industrial revolution a nation of warriors, like the Moguls, could take on big empires around them such as the Chinese and Islamic empires, who possessed more advanced technology and administrative systems than the Moguls, and conquer them. After the industrial revolution, however, technology became the decisive factor in winning wars and dominating nations. Also technology became the decisive factor in influencing other factors of civilization within each society—e.g. science, arts, philosophy, communication and governance.

Yet the industrial revolution was not a spark of magic. It began small and picked up pace. It required certain historical precedents, discoveries in science and techniques that were contributed from a long line of civilizations in human history to that date. It also required a set of socioeconomic and political conditions that happened to crystalize in Europe at the time, and it took particular combinations of resources, investment and competition to give it the thrust it needed (O'Brien and Williams 2004). Neither did it start with technical marvels right away. The industrial revolution was demystified by historical analyses.

### The pristine West[8]

Narratives of history are stories told. The difference between them and other types of stories is that significant levels of accuracy are claimed in depicting events that had significant impacts in peoples' lives in the past with consequences that may still be felt by some of us nowadays. Basically, if the story has a verified claim to depicting actual events that

---

(8)    This sub-section is largely a reproduction of a published article by the author. See Sheikheldin 2015b.

happened sometime in the past, we tend to name that story 'history'. Being a story, however, it cannot take you fully back in time. It can only deliver to you what the storyteller(s) saw, perceived, and remembered. In other words, not everything. Storytellers can be worse or better than a video camera. They can only see events unfolding from one angle at a time, just like the camera. They cannot see 360 degrees at once. In addition to that, and unlike raw camera footage, storytellers edit the footage by having them processed through their memory and perception before they deliver the story to you. That can make the story richer or poorer. Memory and perception are woven with language, coloured with biases, and transferred to others in wraps of context. The end product we receive is anything but a fully accurate and comprehensive account, and we shall always have that in mind and live with it.

That said, it follows that no one should be surprised that there are many different versions of history out there; as there should be. There may be some dominant versions at given periods in parts of the world, depending on the order of the day, but there has never been a single version of human history since the recording and communication of history – i.e. storytelling – ever began. We can give brief examples to that. Nowadays we live in the era of 'the pristine West'. It is the era when the dominant version of history says that Europe, throughout recorded history, was the centre of historical developments of human societies, not just by influencing other civilizations outside of Europe over time, but also by having a continuous line of historical heritage that is almost autonomous of any major influence from outside of Europe. That continuous line is often depicted as follows: "ancient Greece begat Rome, Rome begat Christian Europe, Christian Europe begat the Renaissance, the Renaissance the Enlightenment, the Enlightenment political democracy and the industrial revolution. Industry, crossed with democracy, in turn yielded the United States, embodying the rights to life, liberty and the pursuit of happiness..."[9]

It is quite a spectacular story, in which the focal point (the hero?) is identified throughout thousands of years, and the outcome is self-evident in today's modern civilization. This story also has another name. It is called the Eurocentric version of history. In this version, the storytellers are obviously Europeans. It is the story of the world according to Europeans. In this version there is no denying that other places around the world had their own histories, or that other old

---

(9)    Eric Wolf, quoted in Hobson 2004, 1.

civilizations interacted with Europeans (such as Egypt and the Islamic empire), or that European colonization and settlement happened to Latin America, Africa and Asia, or that there were already human groups in the Americas and Australia when European settlers arrived there. None of that is denied, but the premise is that all of it was not as critical in shaping human history as was the internal, self-motivated and self-generated dynamics of Europe's history alone.

It should be quite obvious that there are a number of problems with this version of history, which is important, especially if it is presented as the most accurate and verifiable version. For example, we now have abundant evidence regarding how ancient China did many things that together made Europe a dominant force in modern history, before Europe did them! The Chinese invented and used gunpowder first, built large ships that sailed across the big oceans, and had many critical advances in metallurgy and other applied sciences before Europe did. What is more interesting is that the Chinese dynasties of those times did not have the same aspirations in conquering and exploiting the rest of the world with those powerful technologies as Europe later aspired and executed (Ferguson 2011). Naturally, different forms of arts and culture, and ways of viewing the world, developed in China. In other words, China must have its own, different, and equally legitimate version of history. For example, the Chinese drew their own version of the world map, centuries ago, with China at the centre of the world – in other words it was not 'the far east'. Technically speaking, China could be at the centre of the world map. Who decided that Greenwich is the centre anyway?

In addition to China, the region of the Middle East and North Africa easily has its own version of history too. Persia, Egypt, and the Islamic Empire have a continuous line of recorded history with many events, developments and major world influences that do not have a major role for Europe in the picture. The same could be said about the history of Abyssinia (Ethiopia) and some parts of India. All in all, we certainly do not have one human history—we have many human histories.

There are some peoples in this world today, however, that have lost touch with their own historical narratives. Their own historical developments were severely interrupted, disconnected and mutilated. Their technologies, social structures, arts and memories, were emaciated. They are left now with the big task of remembering their own stories. An African proverb says that until the lion learns to speak, the tale will always favour the hunter's side. It should be obvious that if the lion was there, saw what happened, and was part of it, then the lion has another legitimate version of the story. The late South African musician Miriam Makeba once said,

> "The conqueror writes history; they came, they conquered, they write. You don't expect people who came to invade us to write the truth about us. They will always write negative things about us and they have to do that because they have to justify their invasion in all countries."

Edward Said, in *Orientalism*, expressed the same point: "Every single empire in its official discourse has said that it is not like all the others, that its circumstances are special, that it has a mission to enlighten, civilize, bring order and democracy, and that it uses force only as a last resort."

That in short is how many of us, children of post-colonial societies, feel about the 'pristine West' (Eurocentric) version of history. And our own re-imagination of how our lives could be rearranged – with technology and institutions, with philosophy and arts, with values and trajectories – is important for us today to excavate and renovate, to demonstrate to ourselves and to the world. Whence we endeavor on that path, we start to see many things differently.

## Innovation and migration: a gift to the West

Following on the above point about differentiating versions of history (histories) and how that affects the way we look at the present, here is an interesting issue. Our world today is shaped by migration, and that has been the case for the last few centuries. Technological innovation and migration happen to have many critical connections.

We know that modern civilization in North America was made by immigrants who kept coming over generations and took control of the land from its indigenous peoples. We also know that modern Europe accumulated its wealth from colonization and the exploitation of the natural resources and human labour of other continents; a trend that started before, during and after the industrial revolution which could not have objectively happened without such exploitation (even while assuming, for the sake of argument, that the technological and scientific innovation that sparked and established the industrial revolution was independent of the said exploitative economic conditions of the time). These historical realities remain important whenever we seek to analyze modern global economics and politics. Yet these historical facts alone do not relay the full reality. For example, they do not stop some from taking exception to European migration to North America by claiming that Europeans are the ones that championed, and are championing, the modern civilization project that makes North America attractive to other immigrants. Even when speaking of slavery, some entertain the thought – or even vocalize it – that while many immigrants (and slaves) provided the 'muscle' Europeans were always the ones to provide 'the brain' of the industrialization and modernization project in that part of the world, without which the 'muscles' would not have made a significant difference anyway.

These claims extend up to this day. Emigrants from other parts of the world to North America (with exception to ones from Europe) are seen, in some narratives, as the ones that come to benefit from the achievements of early European migrants in 'the new world'. Since North Americans today would like to enjoy the wealth, relative safety and relative liberties – compared to most other parts of the world – it would not make sense for them to keep welcoming new migrants with open arms. Large numbers of people from the North American middle and working classes seem to sincerely believe that accepting more migrants will deteriorate the living conditions of those who already there. Many taxpayers also seem to sincerely believe that their governments will be spending resources on sheltering and feeding the newcomers instead of using those resources for maintaining and improving public services, and also believe that migrants render the local economy in worse shape as they consume more than they produce. Some also argue that even when these migrants actually work they lower standards of work and threaten the job security of "locals" (i.e. older immigrants).

What has been clarified, and needs to be reiterated time and again, is that all these claims, above, are objectively false. Their falsity has been proven by the same data that speak to their concerns. We will reference some of that data below, but in particular we should pay attention to the role of migration in technological innovation. As has been said before "innovations in science and technology are the engines of the 21st century economy."[10] Therefore it matters very much today to ask: who makes this technology? Who innovates technology and science today, and where are they located?

So far there is a legitimate claim that most of those who make, remake and innovate technology and science, are currently concentrated in the Northern hemisphere; more particularly in parts of the world that are

---

(10)  Video on Tech Insider: "Neil deGrasse Tyson has a problem with all the US presidential candidates". October 16, 2015 on: http://www.techinsider.io/neil-degrasse-tyson-politics-presidential-election-debates-2015-10

globally referred to as the West. They mainly include North America, Western Europe and Australia. There are other places around the world today that are big hubs of innovation, such as China and Japan (and we can nowadays also speak of India, Brazil, and the Asian Tigers) but the West still holds the lead. The evidence-based argument we have here is that continuous waves of migration to the West – with focus on data from the USA as an example – positively correlate with the continuous advances in both wealth, on the one hand, and in science, technology and innovation on the other hand. This positive correlation shows, in more detail, that the continuous waves of migration to North America are one major reason for the continuous advances in wealth and, more importantly, science, technology and innovation in that region. Innovations in technology and science in the West are no longer 'Western' in historical-cultural terms, because they are realized by the innovativeness of demographically diverse groups, in general, with particularly large contributions from immigrants.

An elaborate recent study was published in a report by the Brookings Institution on 'Economic Facts about Immigration' (Greenstone and Looney 2010). The study investigated what the data says about the reality of the economic impacts of continuous waves of immigration to the USA. It investigated what immigrants take, when they come to the USA, what they give back to the country, and also where they generally end up in the fabric of American society. The report itself is fully available online, but I'm going to focus in particular on two areas: what immigrants contribute to innovations and advances in science and technology, and what immigrants contribute to the cumulate wealth of the USA.

Consider this: If we put first-generation immigrants on one side, and put the entire USA population on the other side (including second-generation citizens who were born and raised by the first-generation immigrants) we will find that:

- In terms of advanced academic education (assumedly a good indicator of advanced knowledge and skills in contemporary societies), the percentage of doctorate degree holders among immigrants (1.9%) is almost twice that of the entire USA-born population (1.1%). At the level of master's degrees, immigrants and USA-born citizens have a similar percentage. Overall, while only 12% of the USA population are first-generation immigrants, 11% of them hold an advanced degree, "slightly above the fraction of [USA born] Americans with post-college degrees."

- With their advanced degrees, foreign-born university graduates in the USA show stronger indicators of innovation than the rest of the USA population: immigrants are 3 times more likely to file for patents than USA-born Americans. And out of each 10,000 graduate students in American universities, about 1,100 foreign-born are granted patents while less than 400 USA-born are.

- Additionally, and overall, the entrepreneurial spirit – which goes along well with the innovative spirit – of immigrants is also higher than USA-born citizens. Estimates are that about 350 businesses are registered monthly in the USA by immigrants, compared to about 270 by those born in the USA.

The data above shows that new immigrants are leading participants in the innovation-driven economy of the USA. They proliferate on both the knowledge-production side and the business-growing side, and they often do so as leaders, despite their smaller percentage compared to the entire USA population. Furthermore, another study by the Brookings Institution found that over 42% of foreign students in USA colleges and universities, in bachelor or higher degree programs, are enrolled in degree programs of science, technology, engineering and math (STEM), and many others are also studying in the business fields. About 45% of them find jobs within the USA economy and work there after graduation, at least for a while (Ruiz 2014).[11]

Another recent study, published in the Harvard Business Review (Hewelett et al. 2013), on 'How diversity drives innovation' asserts that companies which employ a more diverse workforce "out-innovate and out-perform others" and are more likely to report growth in their market share and in capturing new markets.[12] The explanation for that is simple: groups of more diverse backgrounds have more diverse perspectives and generate more diverse ideas, leading to more innovations. It is also no surprise that the leading cities in the world today in the innovation economy are overall among the most multicultural cities in the world as

---

(11)   Needless to say a good portion of these foreign university STEM graduates later seek to continue living and working in the USA.

(12)   Diversity was measured, in this study, in two dimensions: inherent and acquired attributes, and included both ethnicity and work experience in other countries.

well.[13] The findings from all these studies mentioned consolidate well and give a more comprehensive and confirmed picture.

Even more interesting is when we look at the overall impact of immigration on the public budget of the USA. According to the official statistics consulted by the Brooking Institution, taken from the US Census Bureau, first-generation immigrant households, overall, pay more taxes than the cost of public services all immigrants use. That is not even withstanding how much they contribute to sustaining the national market as consumers of goods and services. The above indicators should also be considered with other realities: that immigration to the USA now is more diverse than it has ever been before—i.e. recent immigrants are coming from more diverse backgrounds than before. Often enough, these recent waves of migration break many stereotypes. For example, a study by the US Census Bureau, between the years 2008-2012, found that "Compared with the overall foreign-born population, the foreign-born from Africa had higher levels of educational attainment" (Gambino et al. 2014).[14] That is, if considered as one group, immigrants from Africa are the most formally educated of all immigrants.

Could we then say that it seems that the USA gains more advances in innovation, knowledge and wealth from the continuous waves of immigration? That the USA would not be in its current high standing, compared to the rest of the world, in terms of the flowing generation of innovations in science and technology, and economic growth, without immigrants? I think we could say that. Indeed, the president of the USA, Barack Obama, said it, on December 2015: "We can never say it often or loudly enough: immigrants and refugees revitalize and renew America."[15]

As knowledge proliferates throughout the world, and as access to information and ideas and tools is more or less evenly distributed worldwide (uneven, yes, yet not in the same fashion as decades ago) there is no surprise that the rest of the world is catching up with the West in terms of advances and innovations in technology and science. At least at the human capacity level—i.e. the knowledge and skills of individuals

---

(13)    Innovation Cities Index 2014, top 20 cities leading the global innovation economy: http://www.innovation-cities.com/media-release-innovation-cities-index-2014-launch/8913#data_tables

(14)    As a friend once expressed about big brain drain migration of highly skilled/educated Africans towards the West, "it seems that even the full half of Africa's cup is also empty." A worrying yet inescapable thought.

(15)    Barack Obama, from a speech on December 15th, 2015 at the National Archives Museum, where immigrants from over 25 countries were sworn in as U.S. citizens.

and groups from all over the world who are fortunate enough to have access to good education and room for experimentation in technology and science. Westerners technically do not lead the world on that front anymore. Indicators show that westerners now have a fair share of that front but do not hold a strong majority in it anymore. Yet, western countries still fairly lead the world on that same front. The secret to that is not really a secret—it is the so-called brain drain migration. The labour market, division of labour and incentives in western countries still have more capacity to recruit and retain stellar minds and creative talents. Bright and fortunate minds from developing societies, who had the chance of an education that matches and sharpens their talents, find it easier to join the ranks of 'first world' societies and build a life for themselves there, since the labour market there can absorb them more efficiently (generally speaking) while the infrastructure and public services (and the wonders of the market) there offer them a quality of life better than that which they would have had to be patient with – at least for a while – in their home countries. Their choice becomes even easier when there is political unrest in their home countries due to corrupt and/or brutal regimes (which might just be supported and/or funded by the same western countries they end up migrating to).

The peculiar thing about migration to the West is that the West clearly benefits from it while at the same time enjoy bragging about it as a goodwill gesture, and at the same time complains about it when it feels like doing so. On this one, the West can have its cake and eat it too.

But there is more to brain-drain migration, as it is not necessarily a good thing for all countries. The counter-narrative is that the current conditions of continuous 'south-to-north' migration is a bad deal for the economic South (i.e. the economic southern hemisphere). It largely means that a big percentage of their bright, innovative, and highly-skilled folks, and potentially important consumers for their growing markets, are lost to industrialized societies that now have more than their fair share of such critical human resources. For example, "In parts of sub-Saharan Africa and Central America, sometimes more than half of all university graduates migrate to OECD countries, with potentially serious consequences for critical sectors such as education, health and engineering."[16]

---

(16)    OECD. "Migration and the Brain Drain Phenomenon". Viewed November 3, 2015 on: http://www.oecd.org/dev/poverty/migrationandthebraindrainphenomenon.htm

In modern history, no country has been able to achieve genuine economic and technological transformation without its own people taking the lead. If this south-to-north migration continues to take place – for understandable reasons but problematic nonetheless – the West will likely continue to get wealthier and more technologically advanced, while most developing societies will continue to deal with shortages of needed qualified human resources.

However, and despite all, there are still some good things happening now in some developing parts of the world. There are big indicators of growing economies, increasing implementation of contemporary technologies, and better education and health indicators in parts of Africa, Latin America and Southeast Asia. Even some funny things are happening: many Westerners are now migrating to developing societies, but in a different way. We can call this form of migration 'opportunity migration'. These opportunity migrants do not need to migrate for economic survival or political reasons, but do so purely for even better wealth and career opportunities. A recent study by the International Organization for Migration (IOM) found that more Europeans currently migrate annually to Latin America and the Caribbean than the other way around (i.e. more than Latin Americans and Caribbean people migrating to Europe).[17] Parts of Africa have also recently witnessed a considerable flux of western migrants that are responding to opportunity calls in some of the growing economies there. Interesting shifts and trends are happening throughout the world.

## Alienation and dispossession

For developing societies nowadays, seeking technological autonomy is a pursuit taking place within a global context of alienation and dispossession. This context, as much as it makes technological autonomy a difficult goal to reach, makes technological autonomy quite necessary to pursue, since it becomes as important politically and socially as it is technologically.

Julius Kambarage Nyerere (1922-1999), the first president of independent, post-colonial Tanzania, was a prominent leader of national liberation in his country, and later an outstanding champion of continental African liberation. As part of his involvement in leading a series of global efforts to study, collaborate and mobilize around the

---

(17)  "More Europeans Migrate to Latin America Than Vice Versa, Study Finds". Global Voices (website), June 26, 2015: https://globalvoices.org/2015/06/26/more-europeans-migrate-to-latin-america-than-vice-versa-study-finds/

development challenges of the economic South, mainly through the Non-Aligned Movement and the South Commission, Nyerere took on a very difficult task of not only understanding and explaining, in clear terms, the particular challenges that are hindering the development of the economic South (which comprises mostly post-colonial societies confined in newly-formed states) but also articulating a strategic way out. In 1986, in a speech at the Nigerian Institute of International Affairs, Lagos, Nyerere articulated particular aspirations of the anti-colonial movements.[18] It was not, he said, that we wanted to simply replace a foreign ruling elite with a native one. If that was what we wanted then we would have had little moral justification to mobilize the masses with us (i.e. the emerging African educated elites). Such change would not have promised them any material transformation in their lives. We pushed for liberation because we wanted a more just distribution of our resources and wealth amongst our people. It was not even mere economic growth that we wanted, for if we only wanted that then we would not have had a strong case to say that we could manage the economic growth of the country better than the colonizers with their experience and competence with the modern ways of running and growing the economy. Nyerere said that when the emerging African elite inherited the rule of post-colonial African states, they inherited "poor, undeveloped, and technically backward" states, which were previously largely used as sources of raw material while most industrialization and supportive modern infrastructure existed in the colonial metropoles. Therefore, Nyerere concluded, the mission of new African governments was not simply to develop their post-colonial societies economically, but also along equitable lines for the masses. The masses were not only dispossessed of the resources that were contributing to modernization, they were also alienated from the process of modernizing Africa. The mission of African national liberation movements was not just to bring prosperity to the native peoples, but prosperity without exploitation. This task however was more complex than initially thought. Pan-Africanism took on an astronomical task, without exaggeration.

(18)   See Nyerere 2011, 52-63.

"History tells us that prosperity is the product of exploitation....It is only the Africans who are demanded to [pursue] prosperity without exploiting anyone. No one knows how to do it this way as no one had done it before....Africa is shouldered with the historical responsibility of providing a case where prosperity is done without exploitation."[19]

Yet, the more problematic point is that many post-colonial leaderships, in Africa and other places, forgot about non-exploitation and equitable distribution of wealth, shortly after political independence was attained. Nyerere noted that, in Tanzania, the leadership opted for a fairness-in-poverty approach, with the aim and hope that the country works together, with other ally countries, to achieve fairness-in-prosperity. It was a more sincere approach, to Nyerere.[20] Yet, as we stand today, we cannot affirm that even that was achieved.

From this perspective we can think more about the interlocked relationship between alienation and dispossession in post-colonial societies. These societies started from a point of alienation from modern technology. They were not given access to master it and and be able to build, use and manage it on their own without the oversight of the former colonial masters. They were also dispossessed of the fruit of their land's wealth. After political independence, it was critical for them to de-alienate themselves from modern technology in order to reap the fruit of their land's wealth. A serious paradox was already waiting for them right after political independence: they could not develop their countries without extracting wealth from their natural resources, and they could not extract that wealth without acquiring the technological capabilities to do so, which in turn also needed some wealth to facilitate. Those post-colonial conditions are not very different for many nations nowadays, and for many reasons.

## Global political economy today

Let us briefly discuss the ways in which politics and power could be seen as relevant to international trade, finance and money, and the degree to which this revolves around the actions of states and international bodies. In other words, let us talk of global political economy. How international policies on trade and finance directly affect the allowable trends and

---

(19)   M. Jalal Hashim, *To Be or Not to Be: Sudan at Crossroads (a pan-African perspective)*. (Manuscript was not yet published when cited here). While this speaks about Africa in particular, it largely applies to other developing societies in others parts of the world as well, such as developing Asia, Central and South America.

(20)   Obviously that statement by Nyerere is only relevant to the time it was said, not necessarily to this date.

modes of international trade and development? These policies are made by the political powers of our age (not by international consensus), and they, in turn, reflect these powers' interest in maintaining and advancing this current global power structure that gives them clear economic advantages over other human groups. We shall try to demonstrate this briefly through two major aspects: international production and international trade laws, and the politics of direct foreign investment (DFI) and development.

The connection between production and trade is consumption. In crude terms, consumption is the incentive for production, and trade is the medium between producers and consumers. The regulation of trade is the regulation of the relations between producers and consumers, but on the international level, the regulation of trade can actually influence both production and consumption, thus influencing, even controlling, general economic relations. An international set of laws of trade can, and does, restrict opportunities of production of certain commodities in one region of the world, while opening them wider in another region at the same time. By doing this, this set of laws creates distinct modes of consumption in each region, due to the availability, price and quality of commodities in each region.

If we agree that the cycle of production and consumption directly and greatly dictates the economic level in any society, then we would surely have a keen interest in looking at the international laws of trade – what means they use and what goals they represent – in order to understand how the political process of creating and enforcing these laws directly affects local economies. The laws of the organizations and agreements supported by global powers, such as the WTO (and its predecessor the General Agreement on Tariffs and Trade) represent the interests of the global political powers-that-be in favouring the economic ideology that gives advantage to the established economic centers of the world: i.e. the liberal market economy. This so happens by practically forcing almost all countries and their industries into a game of competition played by rules that give advantage to the countries that already have the upper hand in terms of capital and technology. In this way, these laws ensure that the status quo usually only changes to the advantage of the global powers-that-be, at the cost of the others, and almost never towards the opposite direction.

Market economy is the ideology adopted by organizations such as the WTO, which promotes its to be embraced by all member-countries by means that seem to be non-compulsory; yet the consequences of not ratifying are so severe that weak countries cannot afford not to ratify (or, more correctly, comply) international/regional trade agreements. This is so because the WTO, and similar institutions, provide a medium of negotiation between less developed countries and the global powers; a medium unavailable otherwise. Less developed countries hope that they can use the room of negotiation they have inside the WTO to make the global powers give them through diplomacy what they cannot attain through self-determination in a globalized world. It happened previously that these less developed countries tried to challenge such international trade laws and they did not succeed (Tandon 2015).

The laws of intellectual property rights are another set of laws adopted under international trade laws that generally assure the maintenance of technological superiority of the global powers. Although they may have some reasonable aspects (such as securing due credit and reasonable compensation for one's innovative efforts with useful outputs), these laws generally disregard the premise that any technology is inherently a social product and only a result of a long line of historically cumulated technologies that built upon each other and were developed in a long line of international exchange of knowledge. There is no major compelling reason to assign rights to control access to these technologies[21] to specific individuals or companies now if it was not for the sake of monopolizing the social benefit of the technology in the name of fair competition, which as has been argued is, logically, unfair.

Many points could also be raised about DFI and development, but the two that we want to briefly address here relate more directly to finance and trade. The first point is about how the global system of trade and finance allows direct foreign investment to take risks by investing in less developed countries, but does not hold investors responsible when these risks come to undesirable results. This so happens because of the great level of capital mobility across national borders that the default global financial system allows – i.e. the ability of foreign investors to pull-out the money that they invested in projects overseas, and across borders, fast when they want to. Such capital mobility makes it easier for direct foreign investment to divert its capital from any country as soon as it feels a threat, causing a great economic imbalance in the country it leaves. The 1997 Asian financial crisis is a good demonstration of this point. Additionally, many DFI and international trade regulations put much

---

(21)    Especially technologies that are more related to vital development sectors, such healthcare, agriculture, energy, etc.

power in the hands of transnational corporations and authoritarian regimes, which contribute to long-term series of labour exploitation, heavy debt, and great dependence on foreign markets and industries. Local industries in the poor regions of the world cannot compete with sophisticated foreign ones that can provide cheaper products and services than they can because of the massive capital investments, infrastructure and advanced manufacturing technologies that support them. Many industries in developing societies have the potential to grow technologically if given the time and investment (a very logical and reasonable demand), but the odds are stacked against them due to international mechanisms such as DFI and 'free market policies'. Such odds did not exist for local industries in Europe, and North America, during the industrial revolution and following it. Local industries took their time to grow qualitatively and quantitatively (the story of the steam engine, which took multiple decades from first invention to becoming an established technology in the British commercial sector, is evidence for that). Yet developing societies, for many reasons, do not have the same relaxed time to achieve technological autonomy. They have to do the same things, more or less, much faster. But how fast? Especially under such constraints.

Pursuing technological autonomy is an evolutionary process that thrives in a self-reliant socioeconomic structure, something which the current international system does not support. These political limitations set by the laws of the global system of trade and finance are directly correlated to the global economy, which makes the challenges of development not only about a society's own capacity in institutions and local industries, but also about a generally hostile global environment (despite how it portrays itself to be the opposite of hostile). The net result perpetuates the vicious cycle of alienation and dispossession.

## ICTs and justice

This section will not attempt to address the various and vast impacts and implicatios of ICTs on developing societies, especially that ICTs are very dynamic and changing in relatively fast pace as we speak. This section will   only address their implications on direct development efforts in the conventional sense. Ursula Franklin (1989) offers a distinction between two types of technology: work-related and control-related. Both distinctions are based on the function of technology in social settings (or how technology is institutionalized in society). The work-related technology aims to improve and ease human output and effort (i.e. production) through technological assistance, while control-related technology aims to increase control over the operation in which it is involved. An example Franklin offers: while a free-standing word processor is a work-related technology, when we link that word processor to a network system through which employees' work can be timed, assignments broken-up, and interactions between operators monitored, we are now dealing with a control-related technology. Perhaps Franklin's choice of example technology – an ICT one – could not have been more proper to demonstrate her argument.

ICTs nowadays have the upper hand in the ease of communication among people and their exchange of ideas. In the development literature ICTs also have the upper hand in terms of how much research has been done recently on their diffusion and success in reaching and changing communities that were not reached in the same quality and quantity by other technologies.[22] While many technologies that serve essential livelihood activities, such as agricultural technologies, are still facing many barriers to adoption in the majority of rural Africa, for example, cell phones (and smart phones) spread widely in a very short period of time. That spread also brought significant changes to these communities. Suddenly the challenges of communicating with the outside world have been significantly minimized. Also, innovative things started to happen. Currently, in East Africa and other parts of the world, mobile payment systems make individuals use money transfer and saving services using only cell phone numbers as financial accounts. One can perform a significant money transaction, for any reason, from one phone number to another in less than a minute. Then the receiver can cash out that money using service booths spread all over the country and in virtually all villages and towns. Additionally, these financial transactions and saving services extend to include companies and institutions, not just individuals. Salaries can now be paid by phone, and items can be purchased. Smart phone applications, in different languages, started to appear to serve particular needs of certain groups, such as farmers, that could use applications, or apps, to follow the prices of their produce in the national markets in urban areas and so on. Changes that together amount to revolutions are steadily taking place in many developing societies in communication frameworks and in means of exchanging information.

---

(22)   See UNDP 2001; Lekoko and Semali 2012; Rensburg et al. 2008; Nasir et al. 2011.

On another level, the proliferation of ICTs has a particular significance for national approaches to infrastructure. Before cell phones, landlines were the primary means for telecommunication between a village and the outside world, as well as between distant villages. The only other alternatives were literally physical (i.e. involved movement of humans or letters for long distances). There was also the use of radio as a one-way communication and information provider (which is still prevalent in many parts of rural areas in developing societies). The resources and time needed to build landline infrastructure to reach an entire country were obviously huge and prohibitive. Now for wireless telecommunication there is still need for infrastructure (such as signal towers and some electricity, as well as extending fiber-optic cables across the land for high-speed internet) but the scale is significantly different.

The other significance of ICTs is that they do not have traditional versions; i.e., there are no customary/traditional ICTs in developing societies that existed before the wave of modernization that came with contemporary technology. In the cases of agricultural technologies, water harvesting systems, masonry, transportation, etc., contemporary/modern technology seek to replace existing native technologies that perform the same functions. It could be argued that this is one of the reasons behind the interesting speedier and broader spread of ICTs compared to the contemporary technologies mentioned above. Since ICTs are basically a fresh sector of technology their adoption process avoids the typical clash with existing technologies and their encompassing socioeconomic institutions, which reduces friction with adoption, since there is less resistance. ICTs tend to be accepted as an entirely new sector of technology instead of a replacement to existing systems that may become too settled to be easily removed or replaced.

With regards to matters of justice, the role of ICTs in developing societies has also been widely studied and highlighted, and it received recognition since before the beginning of this millennium.[23] That is because of the enormous possibilities that come with inherent decentralization of information and communication that comes with ICTs.

"The advanced information systems that are used to improve economic development can also be used for decentralized communication and information access at relatively low cost. New media can be linked together inexpensively and easily to local and global networks offering

---

(23)   See, for example, Kavanaugh 1998; Diamond 2010.

users a potentially vast array of information and communications services; such as electronic mail, small discussion groups, global databases, libraries, electronic newsletters, and conferences. The internet and other computer networks make small group participation and interaction convenient and affordable on a scale that has never existed before. Increases in multipoint-to-multipoint communication have the potential to strengthen democratic processes and diverse sources of information. Control over the communication process by both the sender and receiver is increased by the two-way interactive capability of computers, thereby potentially increasing intellectual diversity and openness...The multipoint-to-multipoint capability extends the scale of interaction way beyond the telephone or letter. Each social critic with network access becomes an informal news editor and publisher." (Kavanaugh 1998, xii).

The possibilities are big, as can be seen, and some of them have already taken effect. In current global affairs, social and sociopolitical movements are heavily present on social media platforms. Sensitive information of all sorts, that relates to peoples and their rights and plights, are becoming harder and harder to hide by central authorities. Education is witnessing qualitative and quantitative transformations as websites and platforms offer education materials, that are culminations of the state-of-the-art technologies, systems and topics of the day, for free (or in some relatively affordable packages) to everyone who has sufficient internet access. Additionally, even new styles of teaching and training in some subjects, such as mathematics and natural sciences, have appeared (such as the Khan Academy model of teaching which is master-oriented rather than the typical class/term progression model).[24] Data and information about many things that used to be arduous to acquire are now at many people's fingertips. Computing technology and devices, along with telecommunication technology, provides a whole system of interactive learning, where people in remote areas can communicate, unify their visions of education and have equal access to the sources of information in all their fields of study, whether in social sciences, cultural learning, or science and technology, for all the levels of comprehension. While there is still clearly unequal access to all that ICTs can offer around the world, due to other challenges, such as access

---

(24)   The author is one of the many users of Khan Academy, which is useful whether one wants to learn a new subject or refresh previously acquired knowledge and skills. From experience, and in agreement with many others, it could be said that this particular platform represents a revolutionary perspective in math and science education by itself. It will be interesting to watch how this model evolves or expands in the near future.

to electricity, skills, time, internet-popular languages, and resources to acquire and use ICTs, big steps on the side of making information accessible are taking place.

For developing societies overall, the core of the opportunities and challenges that come with ICTs can be summarized in two words: education and democratization. In other words access to learning and access to expression. For both, ICTs do not provide direction. They only provide content and platforms. They are an advanced medium of exchange. That is why ICTs are double-edged swords. Without the existence of national strategies in fostering education and democratization then ICTs can be used for counterproductive streams. As we speak, ICTs are being used by some authorities and private organizations to invade users' privacies, monitor their activities without consent, and block their access to particular information deemed sensitive or 'classified' by those authorities. On the other hand, many individuals and small groups use ICTs for a variety of unproductive or counterproductive endeavours. Terrorist organizations use the internet as recruitment grounds. Hate speech, abusive practices and spreading of inaccurate and false information are rife in cyberspace. While we could clearly dwell on that subject, it is not of significant relevance to our main topic, and clearly, as a medium, ICTs are capable of being used for education and democratization, and will continue to be ripe for productive, and very productive, use.

The main take-home from this discussion of ICTs is that, as a revolutionary medium, ICTs can be utilized to serve goals of equitable and sustainable development in very effective and efficient ways. Yet they do not serve those goals by their mere existence. Engaging ICTs for development should commonly be a part of more comprehensive and coherent development agenda and strategies for education and democratization.

## Justice, the factory, and cooperatives

The factory is one of the most overwhelming inventions of the industrial revolution. That is because it is an institutional creation that was, and still is, a direct response to modern technological transformation.

The need to gather all workers under one roof, to function together in a production and assembly line with respective machines in a way that makes them too – the workers – part of a great machine (which is the factory) happened because of particular inventions that were introduced to textile production in Britain in the wake of the industrial revolution.

These inventions were the steam engine and inventions that were able to speed up the process of spinning.[25] Before that, workers used to produce their quota of spun threads at their own homes then deliver them to their employers on schedule (often weekly). With that arrangement it was clear that spinning was the bottleneck of the production process, and it typically took much longer than the weaving process. When inventions that were capable of spinning more than one thread at once came into existence, the idea of running the machine with the steam engine was soon to follow. Consequently, with the new way of spinning threads it became intuitive to get the workers under one roof rather than the old arrangement of each one working from their own home, since the machines cannot be divided into households.

Enter the factory.

It was not an overnight change, and it was not easily accepted by the workers. The whole transformation process took some stressful time for all involved, and female and child industrial labours were involved (O'Brien and Williams 2004). Yet when it was finally established it soon proved superior in both quantity of production and quality control. When the idea of the factory gradually transferred to other industries it proved to be even more intuitive with heavy and more complex industries.

Today the factory institution is one of the highest signifiers of industrialization. It is even quite often the poster symbol of the industrial revolution and the post-industrial revolution era. There is no denying, overall, that the emergence of the factory was by all standards quite a breakthrough in industrial relations—be they human-machine relations or worker-employer relations. Even more so, the factory influenced entire social arrangements and redefined them. The workplace and work environment became very different concepts and experiences. Most importantly, the factory ushered in a new age of industrial life, with complex technologies and mass production, without which modernization could not have taken place, and perhaps many other aspects of modern life we do not often connect with the factory—such as advanced R&D processes, scientific discoveries and innovations, technical communication, and of course the marvelous spectrum of personal and high-tech products we cherish in our lives nowadays.

Yet there is a persistent problematic characteristic of the factory. It is generally a notoriously exploitative place. The physical and intellectual

---

(25)    Spinning and weaving were the main processes in textile production.

work of many renders wealth for the few who own the machinery with which the many produce, while the many typically share none of the fruit of their labour; only wages determined by the few owners of capital. Also, relations are very hierarchical in the workplace that is the factory. Those who produce do not participate in the decision making that guides their production, and those who do not produce get to decide and enforce their decisions upon those who produce. That is the general picture of the day, with some few exceptions and some few variations. During tough times, when production and consumption are not in harmony, the first who suffer, and the ones whose livelihoods are most threatened, are the workers, while the managers and employers seek to minimize their losses (as the meaning of losses can sometimes literally be a decrease in profit). In addition to that, in modern factories, the more advanced they are the more we find that the most disposable and replaceable element in the production line is the human element; particularly the floor labourers, technicians and machinists. All around the world, the worst conditions of work are found to be in exploitative factories in developing countries, called sweatshops, where some of the most inhumane conditions are witnessed and tolerated.

Evidently the factory, in general, is not a place where social justice thrives today. Therefore, it is not a sustainable industrial-social model. While it is accurate to say that the factory is not intrinsically good or evil, but rather what we make of it that renders it so, we still must emphasize that something must be done about the factory institution to make it more consistent with human ideals of justice and prosperity for all.

Enter the cooperative model.

While one cannot claim that the cooperative model is the only viable model for transforming the factory into a more just and humane workplace, one can still claim that it is a highly viable one. It is therefore worth the utmost consideration and promotion.

The origins of the cooperative movement go back as early as the mid-1800s. Cooperatives have been established throughout the world – in developed and developing societies – in various industries. Nowadays cooperatives have international support and established international networks, such as the International Cooperative Alliance.[26] Recently the year 2012 was declared 'the international year of cooperatives' by the United Nations.[27]

---

(26)   International Cooperative Alliance website: http://ica.coop/en
(27)   International Year of Cooperatives 2012: http://social.un.org/coopsyear/

"Robert Owen (1771-1858) fathered the cooperative movement. A Welshman who made his fortune in the cotton trade, Owen believed in putting his workers in a good environment with access to education for themselves and their children. These ideas were put into effect successfully in the cotton mills of New Lanark, Scotland. It was here that the first cooperative store was opened." (Alter 2007, 3).

The size and impact of cooperatives in today's world is quite big (see Table 3). Equally, the size of literature on cooperatives – scholarly, policy, historical, technical, etc. – is visibly large and respected today.[28] Furthermore, cooperatives themselves are categorized into several models and operative schemes, such as producer cooperatives, consumer cooperatives, hybrid cooperatives, and cooperative networks.[29] There are industrial cooperatives, agricultural cooperatives, and service cooperatives. There are also state-sponsored cooperatives and private cooperatives. There are also many schools of economic thought that incorporate the role of cooperatives in their analyses and proposals.[30] There are also public policy studies and frameworks that deal with cooperatives as a viable field of alternative economics to the prevalent models in market economies.[31]

The main premise of cooperatives, and how they started, is member self-management. Cooperatives apply democratic approaches to sharing surplus and human resource management. Instead of the typical market economy model of 'capital hires labour' for generating profit, cooperatives work on the principle of 'labour hires capital' for purposes determined by labour (Spreckley 1981). Collective ownership and decision making of businesses, by those who operate the business or use its products/services, replaces the dichotomy of owners/employers vs. workers/users. Even when sufficient capital is lacking among the collective group to establish and maintain their business they 'hire' capital by borrowing or inviting investment with the arrangement that returns on investment will be paid later while the ownership and management of the business remains with the collective. In this model, room for exploitation is minimized, for those who have access to the fruit of their labour, and who direct it, are the labourers themselves. Consumer cooperatives apply the same principle but with emphasis on the collective as the users/consumers of the product. In both variations

---

(28)   See, for example Granados et al. 2011; Alter 2007; Ridley-Duff 2009
(29)   See Rodgers 2008; Mair and Scheon 2007; Dacanay 2012
(30)   See Jossa 2005 and 2009; Gilbert 2009; Baldacchino 1990
(31)   See for example the journal *Annals of Public and Cooperative Economics*

the shareholders are those with direct relationship and livelihood investment in the business.

If we take a look at one big example of an industrial cooperative, we can talk about The Mondragon Cooperatives of the Basque region of Spain as a very good and successful example of a cluster of technology-oriented cooperatives. Mondragon cooperatives are formed as a group of corporate cooperatives – including industrial ones that produce modern technological products for the construction, high-tech machinery and home appliance sectors – that exist in proximity to each other and usually operate as a community, sharing multiple resources, infrastructure, logistics, R&D and financial services (Mair& Schoen 2007). Mondragon cooperatives operate in a cluster model that can be shown to have served them well and also served the efficiency of their global supply chain. Currently, Mondragon has a total of 260 businesses and cooperatives working under its umbrella, with over 74,000 employees - the majority of whom are members/owners – and over €12-billion of total revenue.[32] Mondragon also has a wide international presence.

In principle, the cooperative model is compatible with many macroeconomic structures. Cooperatives operate in capitalist market economies as well as in socialist economies. They are also widely present in countries that adopt a mixed-economy approach. In some countries they are heavily associated – and sometimes controlled – by the state, while in other countries they are more independent.

Can the cooperative model change exploitative conditions that are currently prevalent in factories worldwide? My own considered guess is yes, they can, if given the opportunity. They have proven able and effective without necessarily sacrificing productivity, discipline and other good business practices. Transforming factories to cooperative workplaces can take various paths, some are gradual and some fast-paced. Even introducing some of the main features of cooperatives into non-cooperative businesses – such as worker participation in decision making, organizing and collective bargaining rights, job tenures, facilitation of buying shares, etc. – makes a difference. In an industrial setting the relation between humans and machines is a high indicator and reflection of the larger relation between humans and technology in society overall.

---

(32)    Numbers taken directly from the official website of Mondragon Corporation: http://www.mondragon-corporation.com/eng/ (Last visited November 7, 2016).

## Table 3: Interesting facts about cooperatives

| | |
|---|---|
| Size of the Global Cooperative movement (to date) | - Cooperatives employ over 100 million people (more than multinational corporations).<br>- Over 800 million members worldwide.<br>- 50% of global agricultural produce is marketed via co-operatives. |
| Worth in global economy | The top 300 global cooperatives have a combined turnover of US$1.1 trillion. |
| Presence | - Cooperatives of various sizes exist in all regions of the world (low, middle and high income societies) and in virtually all socioeconomic sectors (agricultural, industrial, services, financing, banking, etc.)<br>- Financial cooperatives are estimated to employ 950,000 globally serving over 850 million people, or 13% of the world's population. |

Source: data compiled from a UK Department of International Development (DFID) 2010 report.

## Technology and poverty

In this last and brief section of the chapter there will be no demonstration of statistics about poverty and wealth inequality in the world today. There are no tables tallying the numbers of people that live on less than one dollar a day, who owns most of the land, what the demographic variations in poverty levels are, and so on. There is no particular talk about malnutrition, dispossession, disease, illiteracy, infant mortality, and so on. Facts about the current human condition of global poverty are easily accessible these days, from multiple reliable sources, for all. There is no need to reiterate them here. However, we shall think about them for a moment, with technology in mind.

The question "how can there be so much wealth in the world and yet so much poverty?" does not seem to strike many people, and there are reasons for that. Poverty always existed throughout human history, many will say. Nothing is new. But what is significant about poverty in the post-colonial era can be summarized in two points:

1. The sheer contrast between the great wealth of some countries and severe poverty of others,[33] highly correlated with geographic regions of the world (economic North and South), and highly related to their roles in colonization history—i.e. former colonizers tend to be rich, former colonized tend to be poor. This contrast shows that a wealth of natural resources is irrelevant; actually most countries of the world that are considered rich in natural resources are among the poor countries. This contrast is increasing with time, not decreasing.

2. The scale and sophistication of production, be it production of food and basic-demand commodities or that of non-basic-demand commodities, which ultimately contribute to a level of material wealth never before achieved in human history, along with a level of integration in the global economy never witnessed before. Yet still, extreme poverty persists along the lines mentioned in point 1 (as well as among the classes of the rich countries themselves).

As we know, our relationship with technology, as humans, started a long time ago. The motivations to improve upon our technologies persist— more convenience, more effectiveness, more comfort, more fluency, more speed in finishing tasks, more joy, more health, more knowledge, etc. While it is not always clear, in the midst of living, whether our most state-of-the-art technologies achieve these goals for us, we are generally in silent consensus on that assumption that we now have more convenience in general with modern technology.

But from time to time it is a good practice to pause and reflect. Is there a visible correlation today between technological advances and alleviation of human suffering relatively, say? Actually, is there any correlation at all? While we can generally agree that there is nothing intrinsically "anti-suffering" about human technology (witness the persistence of war machinery) we still cannot deny that we seek to advance in technological achievements so that we do the things we want to do more conveniently and effectively. And one of those things we all claim to want to do, as humanity, is to alleviate human suffering for all.

So the legitimate question that comes to mind is: what are we really seeking, as humans, to achieve with our advances in technology? And if we claim that our aspirations are more or less united when it comes to

---

(33)  Wealth is defined as the value of physical and financial assets minus debts.

ending poverty all over the globe, how come our advances in technology do not seem to have been consistent with that aspiration?

Indeed some may claim that humanity has come a long way, overall, in improving life conditions for people across the globe. We can recount the battles humanity won against many diseases using modern medicine and improved technologies of hygiene and food security. Others claim that more of the global population nowadays has access to security, knowledge and convenience than ever before in history. Yet, there remains a disturbing reality today that not some, but all, cannot deny, and that is the stark disparity in the human condition on earth. There is stark disparity in achievement of fulfilling the basic needs of a dignified human life, disparity in access to equal treatment and means of self-actualization, disparity in safety of lives and futures of children, and disparity in accessing the fruits of human knowledge and work. It is a situation far from just, and suffering is rife, with no visible signs of ending soon if we go with the business-as-usual approach.

Something must be done.

Chapter Six

# Selected Stories

*"Technology is not a noun. It's a verb."*

–Stephen Fry[1]

In this chapter I expound a little more on some topics within the bigger umbrella of the technology, institutions and development discussion. They can be considered case studies, further discussions, or applications of what was formulated in the previous chapters to some real world historical phenomena. I simply prefer to call them selected stories. Most of the content of this chapter is taken from previous publications by the author that were modified and updated. The various case studies and topics presented highlight the complexities of technology-institutional dynamics in contexts of development.

## How appropriate is 'appropriate' technology?

The technological divide in the world today is deep and multi-faceted, as explored in the previous chapters. It persists in all aspects of livelihoods, and permeates all activities of creating, using, maintaining, learning and communicating. One of the complications this begets the process of technological change is that, in most sectors and applications, technological change has to go through phases. On a visualized spectrum, one side of the spectrum has customary/traditional and

---

(1)    "The Future of Humanity and Technology", 3 October 2017, Shannon Luminary Lecture Series.

small-scale technology, and the other side has modern and large-scale technology.

Miscalculated shortcuts in the technological change process can be, ironically, a waste of time and energy. At other times it can be quite dangerous. The reason is that when a technology is introduced without being properly embedded in the local socioeconomic institutions, the results are quite often the opposite of progress and sustainability. Appropriate technology is one of the approaches that are concerned with going through that technological change spectrum carefully, choosing what, when, where and how to introduce desired change consistent with the local capacities of developing communities.

## The Concept of Appropriate Technology

As mentioned earlier, appropriate technology is generally defined as technology envisioned with a specific development context in mind. It seeks to be sustainable for that context in particular—i.e., locally manageable in terms of know-how and material resources.

> "Compared to conventional technologies, appropriate technologies typically are less capital intensive; more labor intensive; less dependent on scarce foreign exchange for imported goods; and easier to maintain, operate and repair. Yet, appropriate technologies are labor saving in comparison to traditional methods of production." (Hyman 1987, 36).

Essentially, appropriate technology performs a particular balancing act between the capital and labour demands of both customary/traditional and modern technologies. According to the conditions of the context, it tunes the capital demands below that of modern technology in order to reach affordability, and it tunes labour demands above that of customary technology in order to reach more effectiveness, efficiency and convenience.

For a while, in its earlier years of global advocacy, appropriate technology used to be called 'intermediate technology'. The terminology was coined by the economist Fritz Schumacher, who wrote the book *Small is Beautiful* (1973) and co-founded the first International NGO that promoted this type of technology—Intermediate Technology Development Group, or ITDG:

> "If we define the level of technology in terms of 'equipment cost per workplace', we can call the indigenous technology of a typical developing country – symbolically speaking – a £1-technology while that of the developed countries could be called a £1,000-technology. The

gap between these two technologies is so enormous that a transition from the one to the other is simply impossible. In fact, the current attempt of the developing countries to infiltrate the £1,000-technology into their economies inevitably kills off the £1-technology at an alarming rate, destroying traditional workplaces much faster than modern workplaces can be created, and thus leaves the poor in a more desperate and helpless position than ever before. If effective help is to be brought to those who need it most, a technology is required which would range in some intermediate position between the £1-technology and the £1,000-technology. Let us call it – again symbolically – a £100-technology.

Such an intermediate technology would be immensely more productive than the indigenous technology (which is often in a condition of decay), but it would also be immensely cheaper than the sophisticated, highly capital-intensive technology of modern industry. At such a level of capitalisation, very large numbers of workplaces could be created within a fairly short time; and the creation of such workplaces would be 'within reach' for the more enterprising minority within the district, not only in financial terms but also in terms of their education, aptitude, organising skill, and so forth." (Schumacher 1973, 150).

The approach of appropriate technology has not deviated, on the technical level, over years, from the description in the excerpt above from Schumacher. The examples of appropriate technology projects and products one can see in developing societies are many, such as:

- Improved designs of local agricultural tools through stronger material (e.g. metal instead of wood for example) or added mechanisms. Also introducing smaller and human-powered versions of motorized modern machinery to replace more labour-intensive tools.
- Brick-making machines that do not need motorization and can be fed manually but produce better quality and standardized sizes of bricks than the local customary technique.
- Cooking stoves made from local materials that use energy more efficiently (especially biomass fuel) and reduce burning smoke emissions, to replace some local methods of cooking that meet less criteria.
- Water pumps, harvesters, carriers and purifiers that improve access to cleaner water and its storage for some communities that have less accessibility to clean water on a regular basis throughout the year.
- Other innovations or inventions that seek to meet particular demands of targeted communities, whether for production or consumption, in a way that is perceived as qualitatively and quantitatively better

than the current local way of doing things, while also not quite the aspiration given the existing, yet unattainable, more complex solutions.

Make no mistake, appropriate technology does not simply mean 'lower price and lower quality'. While many appropriate technology products indeed seek to achieve more affordability and in so doing compromise the optimum level of quality or durability that is theoretically attainable, it is not by such criteria that an appropriate technology is simply categorized as so. What puts the 'appropriate' in appropriate technology is its ability to satisfy conditions of 'local context satisfaction'; of being harmonized with the existing limitations and capacities of the local conditions (i.e. resources, skills, market, demands, etc.). A cheap and simple quality product that is imported from somewhere else is not likely an appropriate technology. Also, when some local producers make, for example, a lower price and lower quality product or machine that exists somewhere else, that is also not necessarily an appropriate technology, albeit a localized technology (which could be a good thing even if not optimum quality yet). Appropriate technology holds no monopoly over 'good' and localized technology for local demands. It is just one approach within the myriad of many that can exist to meet technological development milestones in the path towards technological autonomy.

For example, Roy (2002) tells the story of how customary handloom weaving technology in India has a unique record of surviving the new mechanized 'power' looms. The general expectation from such 'competition' was the gradual disappearance of the so-called traditional technology, due to conventional assumptions about productivity and market factors, but the reality in south Asia is that handloom weaving not only survived, but increased productivity and sustained a market for itself despite the flourishing of power weaving on the other hand. Moreover, and quite importantly, technical improvements in the designs of customary handlooms took place steadily in certain periods. Handloom technology was able to survive and gradually evolve because it continued to satisfy the local context conditions for an appreciable size of the population.

The conception of appropriate technology could be fairly attributed to many innovators and leading thinkers on development in the post-colonial era; notably from the economic South itself. As early as 1958, Mahmoud M. Taha, a prominent Sudanese national and anti-colonial figure, authentic theologian, civil engineer and sociopolitical leader, wrote about the prospects of changing the formal education system of independent Sudan (two years after Sudan's political independence):

> "I believe that if we want for our education to be beneficial and fruitful, it is indispensable to refer to foundation assets in planning the educational curricula. A quick glance at these assets tells us that those educational curricula must aim first to teach a student how to educate themselves and how to be obliged to continue that education throughout their lifetime (self-education or self-learning). Filling the students' minds with a selected set of diverse and numerous bits and pieces of information, on which they have to pass exams to be able to proceed for further stages, is a useless – and in fact, a harmful – method…We cannot prepare a person for life as properly needed, unless we are able to sustain them – within the regular schooling systems that civilized governments provide – with the scientific style by which they can continue educating themselves. The benefit of education rests on the ability of the learner to adapt to their environment."[2]

Taha proposed some practical guidelines and measures for a national education system that was appropriate for the Sudanese environment, culture and economy. Changing the education curricula – which was inherited from the colonial system – was his central proposal. He argued that the colonial education system, as a whole, was not a system that was designed according to the Sudanese context. It was rather designed according to the British experience and interests, which historically was not aligned with Sudanese experience and interests. When they introduced formal education to Sudan, the British did not deploy a national development curricula for Sudan, but a curricula that aided their own agenda of using the colony of Sudan as a resource for the British Empire, and as a subsidiary of it. It should be obvious that colonial powers would not introduce formal education curricula that suited independent countries. Consequently, that country would need to revise that curricula to be consistent with the new conditions post independence.

Appropriate education has to be appropriate for its context, Taha argued. He proposed the change of the educational stages, and the inclusion of technical education from an earlier age in order to teach students to be productive and well-utilizing of their settings rather than filling their brains only with theoretical material that they can

---

(2)   (Taha 1958) The excerpt is taken from a translation (to English from Arabic) on the website www.alfikra.org. I made some changes to the translation as I saw fit and compared with the original Arabic text.

hardly connect to their daily lives. Community participation in the construction, organization and maintenance of schools was very important, he proposed. He advocated that schools' premises can be built by members of the communities themselves, especially ones that are built for early education levels in rural areas, using the same materials, designs and techniques they build their houses with. The state gets involved with the people, he argued, in the supervision of schools, but neither of the two monopolizes that job. He also proposed that higher education institutions, such as universities, are not as essential as covering primary and secondary education for all, because the country first needed to work within its economic limitations to cover the basics for the majority of the population, rather than provide quite advanced levels of education for a select minority.[3] "Coverage must come before perfection."[4] That was besides the general reality, Taha argued, that university education in low-income countries mostly creates an elitist group that generally does not contribute efficiently to the growth of their nations (in proportion to how much their nations invested in them). Only a few of those university graduates later prove to have been a worthy investment by the country.

> "Male and female [university] graduates constitute a new class that is not concerned much with the poverty conditions that the nation suffers. Moreover, they are surely incapable of living in villages and rural areas to be able to understand the reality of their society, and thereafter, foster rural development." (Taha 1958).

That general critique of university graduates of newly independent countries proved to be generally accurate in many cases. Coulson (1982) tells a similar tale of the elitism that was cultivated by the early graduates of the University of Dar es Salaam, Tanzania, and how a large number of them grew to see themselves as deserving of exceptional socioeconomic treatment rather than a national investment that was to be returned in dedicated labour for the interests of the masses. The early graduates of the University of Dar es Salaam were by no means an anomaly in the general story of early university graduates in newly independent and poor countries.

---

(3)    By the standards of the time at which these ideas were proposed, university education was indeed quite an advanced attainment, costly for the country (to accommodate university students) and it stressed its resources disproportionately for a small number of students out of a large population.

(4)    Ardant 1963.

Taha also argued that local technologies can be comprehended, standardized and improved in developing countries faster than the intensive theoretical education foundation that was often provided in modern colonial education curricula before any hands-on training took place. Since local technologies are less complex – at least to the natives – than the ones in high-income countries, and are more understandable by direct practical training than by extensive theoretical training, it was a better approach to teach these technologies earlier to native school students.

Taha's arguments, formulated since 1958, were visible in the general idea and principles of appropriate technology that had been vocalized when independence movements began articulating themselves. Julius Nyerere in Tanganyika (later Tanzania) spoke and wrote along similar lines as Taha (as we will see later in this chapter) as did Gandhi who had proposed similar measures for sustaining and supporting rural development in India. Amilcar Cabral, an agricultural engineer by training, during his leadership of Cape Verde and Guinea Bissau's anti-colonial liberation movement, proposed new ways of managing and implementing techniques and plans in the agricultural sector in West Africa.

So, generally speaking, the conceptualization of the appropriate technology approach was no strange or novel idea to a good number of leading national liberation luminaries. What appropriate technology meant was a measure of establishing self-reliance, and growing along that line to reach a nation's own modernization on the technological level—i.e. technological autonomy. To these luminaries, appropriate technology was a sociopolitical strategy as much as it was an economic and technical one. More or less it was also the same approach to appropriate technology that Schumacher (1973) took in his famous book *Small is Beautiful* (which is accredited with making intermediate/ appropriate technology a global development field). Yet, the many international development agencies and NGOs that took strong interest in promoting appropriate technologies, from the 1980s up to the present, rarely seem to emphasize that sociopolitical aspect. What appropriate technology means today, to most, is not necessarily what it meant to the earlier advocates.

## Appropriate technology vs. aid and relief

Following decades of popularization of the appropriate technology concept in the international development arena, one cannot say that the anticipated results have been realized overall. Yet the idea and trajectory of appropriate technology, promoted by international development organizations (especially NGOs), does not show visible signs of abatement. Perhaps for reasons related to the core principles of appropriate technology – sustainability, local capacity building, and self-reliance – that continue to be valid, if not more, after those decades. As long as appropriate technology projects remain consistent with the principles of self-reliance and local capacity building, they play an important strategic role in provoking the forward trajectory of technological autonomy. However, little is useful and sustainable from these projects when they lose consistency with these principles. Nowadays, ironically, a majority of those designated as 'experts' of appropriate technology happen to be Westerners, from countries where they themselves do not use appropriate technologies in their daily lives. There is something wrong with this picture. Appropriate technology is treated in simple material terms, similar to international aid and relief— the local beneficiaries are mostly recipients, and some of them at best assistants, not leaders and innovators. But when you get the chance to see locals as leaders and innovators of appropriate technology projects, you can witness a moving trajectory towards self-reliance. That is when you sense a genuinely fruitful and sustainable appropriate technology approach at work.

In summary, the problem with contemporary appropriate technology efforts is that too many – perhaps a majority – of the schemes that claim the name reinforce dependency, because rather than focusing on unlocking and establishing technological aptitudes among local communities, they have become about mistaking the proliferation of somewhat affordable consumer products with sustainable technology diffusion and localization. The philosophical underpinnings of appropriate technology were sound and it was historically conscious in approach, whereas, currently, the present efforts have largely been co-opted by trends that do not represent those philosophical underpinnings and are quite ahistorical. This needs to change.

## Dams and development

As articulated by Assefa and Frostell (2005), "Implementation of long-lasting, expert-driven solutions, such as new technical systems, require acceptance by the public. In other words, public (social) acceptance shortens the time between the first discussions of new technical systems and their implementation." There are economic and political imperatives for consulting the public with regards to big technological schemes, especially when the claim is that the beneficiaries of these schemes are the public. Making engineering decisions on public projects should not be only engineering-oriented. Decisions like these have to consider persuasive justifications capable of being communicated to the public.

That said, it does not mean that engineering decisions always work this way, but it does not make a fallacy of this argument. Decision making about technology is a socioeconomic process as well as it is technical, where the knowledge, perception, and concerns about the impact of this technology on a social level are always present. Yet we need to clearly identify and articulate the methods that make the social impact of these decisions, in the short term and the long term, comprehensive and effective. This is one area where the bridging of technology-oriented and policy-oriented folks is vital (as discussed in chapter three). It is feasible, and important, to consider the social impacts when making engineering decisions about designing and actualizing technology.

Some of the most interesting cases in the developing world where this particular interface – technology and policy – is most in evidence are the cases of dams and the national development agenda; particularly large dams and their impact in terms of inclusive economic potential. Dams, energy and irrigation are a large and controversial topic since before the political independence of developing countries. It is one of those issues where technology and institutions either cooperate or collide. There is a peculiar additional complexity in that we are dealing with strategic national resources such as water. In this section we will look at two relevant stories from different parts of the developing world: the Nile basin and the Indian subcontinent.

## The Nile: dams in Ethiopia, Sudan and Egypt[5]

The stories of the Aswan High dam (Egypt), Merowe dam (Sudan) and the renaissance dam (Ethiopia) are very interrelated. For one, they share the same resource: the Nile. Along with that – and because of that – each of them was/is a technological megaproject that brought in its wake ecological, economic, social and political controversy.

The Nile is widely known as the world's longest river. It originates from and runs through a number of countries in North-East Africa (Burundi, Rwanda, Democratic Republic of Congo, Uganda, Tanzania,

---

(5)    The content of this section is largely borrowed from a previous newspaper article by the author: "Ethiopian Renaissance Dam: Friend or Foe?" *The Citizen Newspaper*, Sudan. June 2, 2013.

Kenya, Ethiopia, Sudan and Egypt). It starts as two distinct rivers, the White Nile and Blue Nile, with the White Nile originating from the great lakes of Central Africa and through Lake Victoria, and the Blue Nile originating from Lake Tana in Ethiopia, until they meet in Khartoum, Sudan, and continue north as one river towards the borders of Egypt, cutting it from south to North, where its delta is; finally ending at the Mediterranean Sea.

The stream of modernization in the region starting from the 20th century had become heavily dependent on consumption of fresh water for several developmental projects inside states, from urban life-style consumption to agriculture to hydro-power demand. Since the first decade of the 20th century conflicts of interest between Egypt and Sudan regarding securing enough water for irrigation, hydropower, human and livestock consumption, besides other uses, was an issue that colonial Britain had to deal with (Crabitès 1912). Britain, which was in control of most of the "countries" it created – with the magic of geopolitical borders – along the Nile, played a key role in establishing the first "international water treaty" about the Nile water. Because Britain had the most interest in Egypt, the so-called treaty clearly favoured Egypt against the future development interests of Sudan and the other countries. In 1902, British authorities constructed the low Aswan Dam on the Nile River, and later raised it twice, to over 100 feet in height, by 1933. From the beginning, the British understood that this move benefited Egypt and not necessarily Sudan, because their investment in Egypt was more important than theirs in Sudan. General Gordon – the British general governor who was killed in Khartoum, the capital of Sudan, by the Sudanese rebelling armies of Mohammed Al-Mahdi[6] – wrote at some point: "The Sudan is a useless possession, ever was so, and ever will be so...I think Her Majesty's Government is fully justified in recommending the evacuation."[7] Other British men of authority were more diplomatic in expressing their opinion about their 'possession' of Sudan, and their clear favouritism of Egypt, by saying that "It was not that Britain loved the Sudan less; it was that she loved Egypt more."[8] The

---

(6)    Mohammed Abdalla, known in history as the Mahdi (of Sudan), a name he gave himself. He and his followers believed that he was an incarnation of an awaited savior in Islamic mythology. While he was considered simply a deranged man in the eyes of the colonial powers he fought, his Sudanese people viewed him, and continue to do, under more various lights: a true saint and savior to some, an early anti-colonial hero others, and still a problematic (and insane) man to others.

(7)    Crabitès 1912, 320.

(8)    Crabitès 1912, 327.

British, with their European colonial partners in East Africa, designed some treaties between the states that shared the Nile and its sources to mainly secure Egypt's lion's share of the river's water.

"Although the Nile Waters Agreement reached in 1929 consisted only of an exchange of notes between the British High Commission in Cairo and the Egyptian Government, it provided for the regulation of the river until the Nile Waters Agreement of 1959...The detailed 1929 arrangements 'appeared to work solely for the benefit of Egypt whose established and historic rights were recognized'. Egypt was assured a minimum of 48 billion cubic metres of water per year, as against 4 billion for the Sudan, and this left approximately 32 billion unallocated. The agreement did not include Ethiopia, and stipulated that 'no works were to be constructed on the Nile or its tributaries or the equatorial lakes, so far as they were under British jurisdiction, which would alter the flows entering Egypt without her prior approval." (Swain 1997, 677).

Nowadays the treaties that were created during colonial times still stir controversy, long after the colonial powers were technically gone, as can be seen in the agreement between the Egyptian and Sudanese governments, mentioned above, in 1959. Bearing in mind that the treaties the British signed in the name of their colonized nations should not legally carry the obligation of being honoured by the newly independent and sovereign countries, this treaty continued, mostly due to the political and military power of Egypt compared to its neighbouring states; a power that was challenged more than once but eventually prevailed. Many political events happened after colonization ended, not necessarily balancing the domination of Egypt. In July 1970, the Aswan High dam was constructed by the Egyptian government with the help of the USSR. The artificial lake created by the High Dam came to be known as Lake Nasser. "Lake Nasser extended 150 kilometres into Sudanese territory, and the Government [of Sudan] was paid 15 million Egyptian pounds in sterling [in the early 1960s during the construction] as compensation for having to resettle as many as 50,000 [Sudanese citizens] who had been displaced."[9] This agreement with Egypt was signed by a Sudanese military government that had clear favouritism for the Arabization politics led by the Egyptian ruling elite at the time. Nonetheless, the political atmosphere of today is significantly different from that of the 1950-60s. International law is more involved in regulating regional water treaties. Recently, other African countries of the Nile basin started to voice their concerns again about the unfair share of water. On May

---

(9)    Swain 1997, 679.

2013, four East African states signed an agreement to seek more water from the Nile, a move that was strongly opposed by Egypt and Sudan. The upstream countries of Uganda, Rwanda, Tanzania and Ethiopia said that the colonial era agreement is unfair and that they want a new deal, but no agreement has been reached with Egypt and Sudan in 13 years of talks.

As for the ecological and technical problems of large dams in general, there is sufficient evidence to assume that any large dam built on the Nile, and its two White and Blue tributaries, would eventually deal with more or less the same repercussions faced by the Aswan High dam downstream. There is no doubt that temporary economic and developmental benefits came to Egypt from the Aswan High dam. However, in the long term, Egypt started to feel the negative impacts of the ecological damage the dam has accumulated. The ecological damage consists of "water logging, salinisation, and river bed erosion."[10] Severe problems occurred from these three occurrences, and can be mainly summarized in 3 points:

1.  By blocking the sediment flowing with the river's water through Egypt, the dam has drastically deteriorated the fertility of the land shoring the Nile in Egypt. Moreover, the sediment compiling behind the dam has also dramatically decreased the efficiency of the dam in generating hydropower and controlling the water reservoir in Lake Nasser.

2.  The delta region in the north of Egypt lost its great fertility and even its level against the sea water, since it was historically created and preserved by the flow of the sediment with the river's water. Today some sources claim that a considerable portion of the power created by the dam goes to efforts to mitigate this particular impact. Another big portion of the dam's power, it is claimed, goes to the production process involved in producing fertilizers for the lands that lost their fertility due to the loss of sediment.[11]

3.  The need for more water in Egypt is increasing, due to the increase of its population, while Lake Nasser is losing its capacity due to the accumulated sediment behind the dam and the evaporation of water from the lake (since the location of the lake is one of the hottest regions in Africa and the world).

---

(10)   Swain 1997, 680.
(11)   El-Khalifa 2007.

Furthermore, there is a twist. It is precisely because of such problems of large dams that eventually Egypt would want to transfer the problem upstream:

> "While Egypt is seriously averse to the idea of any water being diverted upstream for agricultural purposes, it has actively encouraged the exploitation of the Nile's hydropower potential. The dams which were built in the Sudan after the 1959 agreement have been advantageous for Egypt in so far as they have acted as 'siltation basins' that have stopped a considerable quantity of sediments from reaching Lake Nasser... Egypt continues to show interest in the creation of more hydropower upstream, particularly in the Sudan, since this would not decrease the flow of the Nile, and is also proposing that the huge groundwater potential in southern Sudan should be exploited" (Swain 1997, 686).

So, we now have two conflicting interests of big brother Egypt (compared to its neighbours) downstream: (a) that it wants to keep its share of the Nile's water unchanged, but (b) it would encourage dam projects upstream that help reduce the quantity of sediments reaching Lake Nasser. So, in summary, as long as dam projects upstream do not affect Egypt's share of the quantity of the Nile's water, then Egypt does not mind them (more or less). This was the case with the Merowe dam, where the consent and support of Egypt for it is well known although upon full completion in 2009, 800 km downstream from Khartoum, Sudan, the Merowe dam became among the largest hydropower facilities in Africa.[12] Yet, the other countries that share the Nile basin are not agreeing to the Egyptian perspective, because they want both the hydropower and a greater share of the water, for other purposes we mentioned earlier. To these countries, this arrangement is simply unfair.

Yet, generally speaking, and from a technological development perspective, are large dams worth fighting for? In the case of the Grand Ethiopian Renaissance dam – being constructed at the time this manuscript was finalized, and creating visible tension between Ethiopia and Egypt, with Sudan and other basin countries caught in the middle – the claim of Ethiopian authorities is that it is mostly a hydropower generation dam (i.e. less for irrigation and/or water supply). Yet

---

(12)   The height of Merowe dam is reported to be 67 meters, and the artificial lake it is creating 200 km long. "With a surface area of 800 sq.km, the lake will inundate 55 sq.km of irrigated land and 11 sq.km of farmland used for flood recession agriculture" (Teodoro et al. 2006, 3). The main purpose of the dam is hydropower production. The $1.2 billion capital investment on the project is mostly international (China Export Import Bank, the Arab Fund for Economic and Social Development, and the Development Funds of Saudi Arabia, Kuwait, Abu Dhabi, and the Sultanate of Oman) (Teodoro et al. 2006).

being a main upstream country, Ethiopia's ability to control the flow of the river's water through this very large dam scares Egypt, and makes it unable to voluntarily concede to Ethiopia's legal entitlement to construct a dam within its territory for the interest of its national development; the same thing Egypt did decades ago with the Aswan High dam. Yet the main question of development policy here is about the compromises taken by governments for the sake of hydropower, and whether hydropower, in such quantity, is worth these compromises. That question arises because development cannot achieve its goals when one sector's gain is another sector's huge loss. Concerns about sustainable development also arise.

International standards such as the guidelines of the World Commission on Dams (WCD) require that an environmental impact assessment report be contracted with an independent specialized agency before proceeding with the project. The report is required to critically address certain issues regarding the impact of a dam project, which are: (a) social issues, i.e. consequences of people's displacement and resettlement from future flooded area; (b) archaeological issues, resulting from destruction or submerging important archaeological sites or places of high cultural value; and (c) environmental issues, i.e. effects of large scale hydrological alteration of the natural river system with major impacts on the environment and water quality.

Large dams are known to cause long-term social and ecological impacts, and those negative impacts usually catch up with the technical functionality of the dams themselves—i.e. even a dam's productivity are eventually affected negatively because of the accumulating ecological 'side-effects'. Furthermore, large dams are notorious for being opposed by native communities that live in the project areas and who usually end up being forcefully displaced and their lands submerged. This repeated event is by itself a clear violator of the peoples' right to development, if we understand the right to development as, "the right of individuals, groups and peoples to participate in, contribute to, and enjoy continuous economic, social, cultural and political development, in which all human rights and fundamental freedoms can be fully realized"[13]

In the world of technology, there exist alternatives to large dams for hydropower production and water harvesting. These alternatives are usually more sustainable and often more effective in addressing

---

(13)    Article 1.1 of the Declaration on the Right to Development, the United Nations, quoted in
        Zafarullah and Huque 2005, 21.

the direct needs of communities rather than the state's central agenda. These alternatives include a range of decentralized models for solar power, wind power, micro-hydro and water harvesting schemes. The 2000 report of the WCD stipulates that, "large dam projects should only be approved where they demonstrably meet the goal of human development – and that alternatives including decentralised energy schemes should be considered from the start." So, with this in mind, one has to ask: are the recent large dam projects along the Nile, such as the Ethiopian Renaissance dam and Merowe dam in Sudan, really promoting a people-centered development agenda, or are they actually serving the agenda of elite groups who are in control of governments?

On the political and economic fronts, the countries of the Nile basin have the right to optimal utilization of their natural resources equally. Additionally, the so-called historical water agreements were simply colonial agreements, unworthy of honouring today. Countries of the Nile basin can and should work together to devise more equitable and sustainable legal and political frameworks to share the resources of the Nile in good practice. On the technological and ecological fronts, however, a blanket opposition to large dams and loyalty to the alternative approaches of hydro-power and water harvesting should be supported, with only very few and careful exceptions that are supported by strong justifications.

## India's large dams: a story of technology and politics

Despite some visible economic benefits, large dam projects in India have severe social and ecological consequences that many argue, in the long term, exceed the economic benefits. Similar to the case of the Nile basin, there is also much debate regarding whether those economic benefits themselves are sustainable in the long term, and thus whether the cost-benefit summaries ultimately favour dam building. For example, it has been reported that the displaced peoples from some projects outnumber India's war refugees since independence (Visvanathan 2004). Other projects are said to destroy whole crop types in some regions and cause extreme shifts in the diets of their populations. In addition to that, many criticisms have been raised about the distribution of economic benefits resulting from dams—that it has been consistently unjust and favoured privileged groups with wealth and power. However, the Indian official authorities do not show practical change in their pro-dam position. Fierce and organized social and environmental movements, with some support from political leaders, have been opposing the execution of these projects with many tools of civil resistance, at the community, regional, national and transnational levels. Inevitably, these clashes of interests and perspective are also epitomized in conflicts over development administration models, as well as in their supporting planning theories.

Planning for large dams across India started with the colonial British Authorities as it did in Egypt. However, after political independence India soon witnessed a bigger enthusiasm for mega-technological projects, in pursuit of modernization. This enthusiasm reflected in the attitude of official authorities. A most prominent proponent of this pursuit was Jawaharlal Nehru, the first prime minister of India and a national independence hero. In one of his speeches at a dam project site, he said:

> "When I walked around the site, I thought these days, the biggest temple and mosque and gurdwana is the place where man works for the good of mankind. Which place can be greater than the Bhakra Nangal Project, where thousands of men have worked or shed their blood and sweat and laid down their lives as well? Where can be a holier place than this, which can we regard as higher?" (Khagram 2004, 33).

By the 1970s India became among the top five countries in large dam building. By the year 2000 the number of large dams in India had grown, from 300 in 1947, to 4000. Today, India ranks third in the world in dam building, behind China and the USA. Many of these projects found willing external funders, such as the USA, the USSR, and the World Bank. Building large dams continues in India to this day. The vision of development adopted by newly independent India was clearly a "progress-as-economic-growth-originated one… Capital-intensive initiatives and the exploitation of natural resources were not only promoted, they were prioritized."[14] The unchallenged National Congress party, a 'steel frame' bureaucratic system (inherited from the British), with charismatic national leadership and an infant civil society, all contributed to the reality that saw federal Indian authorities able to autonomously plan and pursue their top-down development vision. Furthermore, large dams seemed to have successfully combined the interests of the major classes that supported this vision: rich farmers, industrialists, and professional elites (e.g., civil administrators, engineers and economic advisors). The public media took part in this pursuit enthusiastically, claiming this to have been ultimately consistent with a vision of a promised modern India for all.

---

(14)    Khagram 2004, 35.

Although the National Congress party was generally under the moral guidance of Mahatma Gandhi, and although Jawaharlal Nehru himself was known to be generally in accordance with Gandhi's mentorship overall (although not in total agreement with him), this development vision was not part of Gandhi's plan for India. His assassination shortly after independence left the 'Nehru vision' virtually uncontested.[15] However, even Nehru himself was said to have expressed reflections, in his last days, about how Gandhi's ideas about bottom-up development – starting from rural areas and focusing on establishing sustainable livelihoods before embarking on big modernization ambitions – should have received more consideration, and that much of what Gandhi regretted about the modernist approach proved to be valid.[16] This reflection of old Nehru was to be echoed later by many Indian authority figures, and is part of a wide and continuously growing debate about the state and development policy in India.

While it is true, overall, that between 1951 and 2000, India was able to increase its food productivity four fold, and achieve beyond self-sufficiency, and while it is true that about 96% of the dams in India are mainly irrigation dams, the correlation that proponents of dams draw directly between the two facts is questionable. Pande (2007) says that optimistic estimates attribute 25% of the increase in food productivity to dam-irrigated areas, and yet it is not logical to attribute all the food increase in these areas to the dams alone, for there are other factors that are not related to the existence of dams. "The increase in irrigation coincided with increased uptake of other inputs and technologies, such as high yield [crop] varieties beginning in the 1960s, fertilizer, machinery and multi-cropping." That is why the WCD only attributes 10% of agricultural productivity in India to dams. Dam opponents argue that even the dam irrigated areas had other options for irrigation and would not have been left uncultivated without dams. Prime Minister Indira Gandhi, since her participation in the United Nations Conference on the Human Environment, in 1972, took the position of a public critic of dams due to their negative environmental impacts. Her son and later prime minister, Rajiv Gandhi, said in 1986 that, "We can safely say that no benefit has come to the people [from dams]…We have poured

---

(15)   In retrospect, Nehru can be safely mentioned among the most successful political leaders of the 20th century. His politics emphasized two goals for India: establishment of a strong democratic system and modernizing India's economy. Both goals have been highly and continuously realized.

(16)   See Shah 1995, 365.

money out, the people have nothing back: no irrigation, no increase in production, no help in their daily life."[17] He was expressing the disappointment of the Indian authorities at the big difference between the early optimistic expectations and the actual results after decades of dedication to the same policy. Despite these strong expressions from two heads of Indian governments, the path which Nehru and his supporters laid at the inception of post-colonial India remained steadily dominant among the political powers of the country. India has continued steadily to build more dams since that time until it reached a world top position, only surpassed by China and the USA.

Yet, in India, there are stories that show a significant and fierce resistance from civil society against the state regarding large dams. An example is the Silent Valley project in the state of Kerala. The Silent Valley covers 8950 hectares of Southwest India, primarily in the state of Kerala. The British colonial authorities identified a site there for a hydropower generation dam in 1929. In 1973, when the state of Kerala made the legal public notification of the commencement of a dam project that would be 390-feet high, there appeared to be no public opposition to the proposal. The investigation by the post-colonial authorities started 15 years prior to that date, and even after that it took three more years to decide to resume the building. This kind of delay in execution, common in India, is attributed to reasons of funding and bureaucracy. The shortage of funds and the complexity of the Indian bureaucratic system have delayed many large infrastructural projects for periods that often reached decades. This, in turn, hindered the building of many large dams, even after their building started, and gave opposition movements time to cultivate and mobilize. The proposal of the dam included the clearance of forests in the valley. That was disturbing to some federal officials who tried to address this internally. Around the time environmental awareness was just becoming a serious matter of consideration in India and around the world. That is why writers such as Patel (1995), Khagram (2004) and others claim that the delay of these projects ultimately gave time for the environmental resistance movement to cultivate influence.

The United Nations Conference on the Human Environment, held in Stockholm in 1972, marked a seminal point in the history of the global environmental movement, after which it became visible in the international scene. India was a participant and a result of that was the creation in 1974 of the National Committee on Environmental Planning

---

(17)    Khagram 2004, 60.

and Coordination (NCEPC) within the Indian Department of Science and Technology. The committee was chaired by Prime Minister Indira Gandhi herself as a demonstration of her personal commitment to the issue. The Silent Valley project came to public attention only two years later, and thus it was a good opportunity to put the NCEPC to work and show India and Gandhi's commitment in action. The NCEPC started an investigation on the project, later producing its "Report of the Task Force for the Ecological Planning of the Western Ghats" (the forest proposed to be cleared by the dam project). The report concluded that the project would have various negative ecological impacts, and that it "should be abandoned and the area declared a biosphere reserve."[18] However this federal committee did not have the final say in the matter, and so it also gave 17 recommendations, as safeguard measures, in case the Kerala state and federal authorities decided to continue with the dam building. In response to that the Kerala state authorities decided to hasten the building process, and led a discussion with the federal authorities (with a different prime minister from 1977 to 1980). The state and federal government agreed to adopt the 17 recommendations and continue with the dam building. However, at that point, both local and transnational non-governmental resistance to the project grew stronger. The local Kerala Sastra Sahitya Parishat (KSSP) organization, the transnational World Wildlife Fund and the International Union for the Conservation of Nature did not settle for implementing the NCEPC recommendations, and led wide campaigns to expose the negative impacts of the project and the doubts over its alleged benefits. A petition was sent to the Kerala high court, which in response ordered a two-week stay on construction, to review the case, which was followed by more similar petitions. After two weeks, however, the high court dismissed the case and lifted the stay on construction. A national lobby by the name of Save the Silent Valley Committee became very instrumental in politicizing the issue nationally and lobbying federal politicians to pressure the Kerala government. The politicizing of the risks and uncertainties of this project, and other similar projects, grew to become a prevalent tool of the anti-dam movement in India. This politicization went beyond the precautionary principle[19] that is known and occasionally practiced in the field of public policy. It sometimes became used as a tool of pressuring politicians and/ or mobilizing national and transnational NGO support against the risk-

---

(18)    Khagram 2004, 44.
(19)    The precautionary principle: taking protective actions even when the scientific evidence
        of harm remains uncertain (Raffensperger and Tickner 1999).

taking planners who wished to advance large dams projects, regardless of the perceived or announced benefits and compensation packages for the affected groups. Dwivedi (1998) explores this founded 'risk politics' tool by the anti-dam movement and attributes much of the success of the movement in India to its effectiveness. It is quite apparent that such a tool would have not worked at all if the Indian democratic institutions were not reliable and genuine at a basic level. This is one of the reasons why India, and not China, became the place of the strongest anti-dam movement in the world. China has the greatest number of large dams in the world and one of the least appealing records of social and ecological mitigations for their impacts.

What happened after the high court decision in the Silent Valley case is that the Kerala state started to move aggressively to complete the project. Again, in 1980, Indira Gandhi came back as prime minister, which presented an opportunity for the anti-dam actors to remind her strongly of her earlier commitment. The case eventually grew to one which the federal government, with a wide NGO and public opposition, stood against the Kerala state government. The Kerala government tried to fight back for a while. Another governmental comprehensive assessment was agreed upon, and this time it took three years when it was initially supposed to take only three months. By the end of these three years, the Kerala state government became weakened and frustrated by not being able to secure federal funding, although it still saw the project as very beneficial to one of the poorest regions in the country. The Kerala government tried to appeal to the opposition by declaring the Silent Valley and surrounding forests – with the exception of the area that would be submerged by the dam – a national park. That was the last measure by the Kerala state government, after which the comprehensive report came out and concluded that it was in the national interest to stall the project. Finally, in November 1983, the government of Kerala announced its decision to permanently stop the project and declare the entire Silent Valley area a national park. The unusual success of the anti-dam movement in this case – which is not by any means commonplace – was highly celebrated, nationally and internationally, and it gave an enormous push to the global anti-dam movement.

Rich farmers, industrialists and professional elites are referred to as India's "dominant coalition of proprietary classes" (Khagram 2004, 35). This coalition has a clear vested interest in promoting mega infrastructural projects such as large dams. Historically, these classes have always benefited from such projects in contrast to the majority.

The history of unjust distribution of benefits from dam projects in India is well documented.[20] On the other side, the anti-dams movement consists of a coalition of those negatively affected by the dams (such as displaced communities) and urban-based activists from NGOs (local and international). Their interests are not always aligned, but on big platforms they prefer to work together for their common interest.

In concluding the discussion of the above cases of dams and development, from India and the Nile Valley, we can say conflicts of interests in mega technological development schemes, such as large dams (and similar schemes), are almost inevitable as we have different sections of society with each one viewing its economic interests as in line with national development interests—the syndrome of 'we speak for all the rest' or seeing development from the narrow angle of one's interests. Some sections in developing societies that have more power and influence under the contemporary state apparatus seem to automatically perceive the implementation of such mega technological schemes as synonymous with achieving control over modern technology, which in turn gives an illusion of taking strides towards national technological autonomy. Yet, we should have enough wisdom to know that 'bigger and more complex' is not always synonymous with 'better'. By now the readers may agree that such mega technologies may help sometimes but are no substitute, by themselves, for good planning and implementation. Steps taken on the path of technological autonomy need to be conscious and steady.

## Ujamaa: brilliant vision, dissonant practice[21]

The story of Ujamaa in Tanzania, during the last half of the 20th century, is quite illustrative and educative in many respects. Here we look at a particular aspect: the disconnection between vision and practice (or planning and management) in grand developmental schemes.

Since 1962, the leadership of the Tanganyika African National Union (TANU)[22] started to articulate a philosophy of national development perceived to be appropriate for newly-independent African states. Rural development, in that philosophy, was the backbone of economic

(20)   See for example Pande 2007; Ram 1995; and Kerr 2002.
(21)   This section is mostly excerpted from a journal article by the author, published in 2015 in the Journal of Pan-African Studies.
(22)   TANU led the road towards the independence of Tanganyika (later Tanzania after uniting with Zanzibar). After merging with the Afro-Shirazi Party of Zanzibar in 1977, it was renamed 'Chama Cha Mapunduzi' (Party of the Revolution), or CCM.

development. Ujamaa, which Nyerere coined as a distinct version of Socialism with an African perspective,[23] focused on national self-reliance by means of government leadership, technical support for rural cooperatives and self-managing rural communities, with focus on agricultural production and education. Equity and productivity were central to the Ujamaa philosophy. Many aspects of appropriate technology and participatory development, widely studied in the world today, can be traced back to early writings of Nyerere on Ujamaa.[24] However, whether Ujamaa succeeded in reality in Tanzania, or has been adopted and modified by other developing countries, is debatable.

Due to that most of the attention Ujamaa received was during the years of implementation and the few years after its fading away, it was thus seldom studied as a historical phenomenon. The study of Ujamaa from a historical perspective has the advantage of looking at the story from a teleological angle, where impacts are assessed in comparison to goals years after the scheme officially ended and the dust settled. Another advantage to the historical study is something very difficult to attempt while the story hasn't yet reached a fair ending, and that is to choose and isolate, from within the whole, one or two particular aspects that ran simultaneously with other aspects during the times the scheme was operational.

In development schemes, big or small, we can generally talk of two phases: strategic planning and management. Strategic planning is "a disciplined effort to produce fundamental decisions and actions that shape what an organization (or other entity) is, what it does and why it does it,"[25] while management can be defined as the next phase and complementing step to having a plan. Management in this sense means the organized leadership effort to implement the plan of the development scheme. A development scheme will not succeed if not carefully planned, in terms of conceptual tools and policy guidelines, but also without a realized managerial structure and temporal deliverables it is nothing more than a theoretical endeavour. Distinguishing strategic planning from such concepts renders more clarity in purposes and conceptual tools than planning is likely to possess. As Bryson (2004) notes, strategic planning is no substitute for leadership, so we can see

---

(23)   'Ujamaa' in Kiswahili translates tentatively to 'communality', 'familyhood' or 'communal cooperation/unity' in English. In contemporary Tanzanian and African literature, it has also become a term coined for Nyerere's version of 'African socialism'.

(24)   In October 2009, the UN General Assembly named Nyerere 'a world hero of social justice.'

(25)   Bryson 2004, 15145.

how the role of management – which is organized leadership effort – is important in the next phases of a development scheme. Indeed plans are only effective if management implements them, hence the structure of planning is contingent on the structure of management (Tustian 2004). A plan that is beyond the managerial capacity available is an unrealistic plan, hence un-strategic.

## The Vision

In April 1962, TANU, the ruling party of Tanganyika (currently mainland Tanzania) published a pamphlet written by its president Nyerere, with the name *Ujamaa – The basis of African Socialism*. The pamphlet presented the ideological foundation and policy guidelines for the Ujamaa national development scheme soon to be put in effect. The central argument put forth in the pamphlet is that capitalism is a mode of production alien to the African context because it has evolved out of European history. Moreover, socialist schools of thought, such as Marxism, which emanated from social and economic analyses of the same European history, also contain many elements alien to the African context, also making them unsuitable guiding visions for the development of African nations.[26] The pamphlet proposed that the best way to construct an African path to development was to derive it from the African values that preceded and survived, allegedly, the colonial era. This ideological point of departure was received with mixed responses by the African intelligentsia at the time, among whom a Marxian conceptualization of socialist development was prevalent.[27] Most of these responses were more concerned with discussing the ideological positions upon which Ujamaa was based than the policy itself. Some writings however defended Ujamaa from a universal socialist perspective.[28]

But Ujamaa was more focused on detail and social analysis and not throwing slogans around. It assumed an African set of communal values, which can be summarized as: (a) respect for all individuals in the community and the role each plays in the institutional/cultural division of labour; (b) all the basic goods in the community are held

---

(26)   It was argued that Nyerere was quite influenced by another version of socialism which was formulated in Europe: Fabian socialism (Coulson 1982).

(27)   See, for example, Shivji 2012a and 2012b, James 1977, and Rodney 1972b.

(28)   For example, Walter Rodney (1972b) demonstrates that, although Nyerere identified with the stream of 'African Socialism' known at the time, Ujamaa was clearly different in content and detail from the other proposals that identified with the same stream. While a mention of it has been deemed worthy, we are not particularly concerned here with the controversial and long-lasting debate of 'African Socialism' versus 'Socialism in Africa'.

in common—food, shelter and the other life necessities are owned and maintained by the community and assigned to individuals and families as members of the community; and (c) work is an expectation of every member of the community according to their designation and capacity, for idleness is as alien as capitalism to African traditional community life. With these values in mind, Ujamaa sought to learn from these perceived African traditions to build a modern socialist country. This modern country was also expected to try to overcome some of the negative aspects that traditional African communities embraced. The negative aspects, according to Ujamaa, are twofold: (1) the inequalities women had to endure, and (2) that production overall was at levels of quality and quantity insufficient for a modern state.[29]

The importance of the rural sector for Tanzania was explained in the *Ujamaa* pamphlet: investment in rural development must take priority over other industrialization plans. At the time, 90% of the Tanzanian population depended on agriculture for their livelihoods, and 80% of Tanzania's exports were produced by the agricultural sector.[30] Strengthening and enhancing the agricultural sector was a logical step in a self-reliant path of development. The pamphlet also explained that, since the majority of the population is rural, agricultural expertise is most abundant and industrialization will have to wait for more capital and technical knowledge not yet in local possession.

To begin with the policy guidelines, Nyerere stated that some counterproductive trends had already started to gain ground in Tanzania. As a result of both the colonial trade policy and the African post-colonial conditions, there was a high tendency for investing in cash crops by Tanzanian farmers at the cost of food crops. Cash crops connect the farmers – small and big – to the wider national and international market without the need to cooperate with each other; hence the second problem: more farmers are working individually now than they used to do in the traditional African village. These two problems came together to make the Tanzanian economy incapable of feeding its own people with enough food crops, besides being very vulnerable to external purchasers who control prices. Other problems that Nyerere pointed out include that Tanzania was drifting into a feudal, pre-capitalist society and away from a socialist one. One trend that was noticed was that big farmers, although praised for their active

---

(29)   Nyerere 1968, aka the Arusha Declaration 1967.
(30)   Devries 1978 and Freyhold 1979.

entrepreneurship and increase of national production, had developed the attitude of investing their money in clearing and cultivating more acres and employing other farmers as workers (mostly seasonal) in the process. These hired workers neither shared the generated wealth of the land nor received secure employment conditions. At the time, unused land was abundant in Tanzania to the point that anyone who cleared and planted could assume legal ownership. This trend was rapidly creating a feudal class system that would eventually influence the political system of the country.[31]

However, at that time some cooperatives were initiated and they needed support in improving management skills and commercial machinery. Nyerere reminded that cooperatives themselves do not produce a socialist system, since a group of capitalist farmers can form cooperatives too; cooperatives should be organized and mentored by the state to be in line with the national plan. The sponsoring of Ujamaa villages was translated into policy guidelines that can be summarized as follows:

- Establishing village-based cooperatives in which all working members of the village are members. These cooperatives manage the village farms. They own and market their goods and also manage their village's basic services (water and sanitation, public education and other civil services).

- Division of labour should be arranged by the community in a democratic structure, with women participating as equal partners. Ujamaa villages are to be encouraged by promotion and incentives but not enforced. Distribution of returns from the community farms should be just, simple and easily understood, and continuous investment of portions of these returns in the improvement of the community services is highly encouraged.

- Dependence on heavy machinery to increase productivity was not an appropriate option at that early stage of national development, so reliance will be on hard work with the traditional technologies and some machinery support when economically attainable. Ultimately the goal is to move towards agricultural mechanization.

- The scheme shall start with existing villages and work its way in re-settling all the rural population of Tanzania in villages (instead of

---

(31)   Such demonstrated class analysis of Tanzanian society led some African Marxian scholars, such as Rodney (1972b), to treat Ujamaa as "scientific Socialism", and to absolve it from the naivety and 'idealism' African Socialism was often accused of.

spread-out homesteads and private farms). Communities of kinship can be good starts on the condition that they don't continue to be exclusively based on kinship in the future.

- There was no need to discourage private farms within the Ujamaa villages, as long as they were operated by their own holders and that exploitation of others (wage-labourers) was not involved in the process. An important guiding policy is that the peasants should have no wage-earners among them.[32]
- Rural industries should be steadily encouraged as they emerged as a result of more productivity, population and communal initiatives. The state shall provide technical support.
- The state shall be the main coordinator of village communities, along with representing the communities collectively in international trade (i.e. the sale of their produce as exports).
- The government shall also direct state institutions to provide extra assistance when needed to the villages.
- National infrastructural schemes to support better production and quality of living shall be undertaken by the state according to its budget and capacity. The two most important resources the government shall provide to this scheme are leadership and education.

In these guidelines, many similarities were there to be found with other rural development schemes undertaken by socialist governments around the same historical period in other parts of the world. The collectivization schemes in the USSR and China around the same time operated according to similar principle of the importance of rural production to the national economy (Davies 1980). The USSR and China governments, however, were not impressed by the 'persuasion' option or the cooperative model from the beginning. The scheme of communal villages that later took place in Mozambique was highly influenced by the Ujamaa philosophy and implementations (De Araujo

---

(32)   The term 'peasant' was used consistently in the Arusha Declaration. While the European origins of the term highlight terminological differences between farmers and peasants (usually based on land tenure relations), that may not apply in the Tanzanian context. It is quite possible to conclude that Nyerere, and others, used the term 'peasant' to refer to both 'peasants' and 'smallholder farmers'. One of the speculated reasons is that, during Ujamaa land tenure for farm labourers was a process of the day (in progress, not yet complete), so it was not practically wise to use the term 'farmers' particularly, or distinguish between farmers and peasants in that context. While this speculated reason, and possible others, may explain the use of the term in the Ujamaa context, the reason does not necessarily, completely, resolve the ideological bearings that come with the term.

1985). There were also a number of similarities between Ujamaa and the rural development program led by the Cuban government of the 1960s.

## The practice

In practice, the main phases that the Ujamaa scheme went through took a progression that unfolded as follows:

- State aid was given to villages that established communal farms and committed to certain organizational criteria that will give them the Ujamaa name. Far flung households and small family houses were persuaded to resettle and join each other to create a village. There were early villages that started on the Ujamaa path before the policy came into effect; those were the TANU Youth Settlements, and they received additional support as pioneer communities. The state aid mainly manifested in financial help to start the cultivation of the communal farms and for individuals in the Ujamaa villages who wanted to invest in their private farms. Additionally, infrastructural support was also provided free of charge to some villages (water supply, some equipment, schools, etc.) and public service staff (extension workers) were assigned to villages to provide technical support.

- Although a degree of success was achieved in this way (meaning that the number of Ujamaa villages increased noticeably all over the country), productivity did not show a synonymous net increase, especially in the communal farms. Many individuals in the Ujamaa villages invested more of their time and money in their private farms, using the state support, and only satisfied the minimum Ujamaa requirements by dedicating some of their time to working on the communal farm. Some environmental disasters (mostly droughts) stood as great challenges to some villages in spite of their efforts.

- After a few years of this trend, the government decided to use the coercive power of the state to resettle peasants and create villages, and to put more strict measures of productivity on already-existing ones. By 1973, after large scale operations, two million people (15% of Tanzania's population) were resettled in 5,556 villages across the country (Boeson et al. 1977). Hyden (1980) however puts the number of resettled Tanzanians in the villagization policy at five million, making it the largest resettlement operation in the history of Africa. Coulson (1982) draws a line between this nationwide forced villagization and the Ujamaa period, since villagization– i.e. practiced by many states around the world– is quite distinct from the Ujamaa vision which was more unique and theoretically elegant than villagization.

- Due to the absence of a robust monitoring and evaluation system, besides other managerial defects, corruption increased among the civil servants working on the scheme. Exaggerated reports (to get more state aid), bribes and accumulation of private wealth among government officials and the emergent kulaks – wealthier farmers of larger pieces of land – became noticeable. Although the central government responded with more centralization measures, to curb corruption, alas, corruption and poor management tools prevailed.

- Despite Nyerere's early statements about the importance of keeping foreign aid to a minimum in the path to self-reliance, TANU's government (which became a one-party regime by 1965) began to allow foreign development agencies to come in and take charge of helping some Ujamaa villages through financial support and infrastructural projects. The World Bank, which became heavily involved in the scheme, was biased towards cash crops production for exportation; something that was quite typical of the World Bank at the time.

- Although it is difficult to pin-point a date when the Ujamaa scheme officially collapsed, the many chronic problems that accompanied its implementation, as well as its discouraging economic returns, led to the mass deterioration of the scheme. Fading away, any remaining enthusiasm for the Ujamaa was not defendable by the time Nyerere voluntarily resigned as president in 1985. Tanzania then entered a mixed economy era and succumbed to structural adjustments as pressured by the IMF and the World Bank (again, typical of both agencies at the time).

### What happened?

There was strategic planning in Ujamaa, and there was a management system. However, were the two connected with each other? Did they integrate and adapt according to what was faced in the reality of implementation? There are many arguments explaining where Ujamaa failed and what the reasons were for the scheme not achieving its ambitions. Some of these arguments trace the problems to a benign, unrealistic image of the African traditional rural life that Nyerere portrayed and consequently built Ujamaa's path upon. Many blame the Ujamaa vision for not acknowledging the class conflict reality within Tanzanian society and that the adamant denial of this reality eventually led to Ujamaa's failure.[33] Others defended Ujamaa in this point, as shown earlier.

It was unlikely from the beginning that Ujamaa would fully succeed as it was in the envisioned program, for many reasons. Those reasons were expressed by a sizable literature, but the most important three are: (1) the initial, and later, underestimation of the influence of the external forces of the global market and financial system; (2) the slow comprehension of the TANU/CCM government of the reality of exploitation and class conflict within Tanzanian society at the time; and (3) unforeseen environmental shocks that happened during the critical years of Ujamaa and caused significant failure in yields.[34] Nonetheless, chances for learning, adapting and making incremental goals were there for the taking had there been more adaptation. Schemes that do not meet their original goals do not necessarily fail if they adjust and achieve some of what was initially expected, instead of an 'all or nothing' case. The Ujamaa policy guidelines that were set in the original rural development plan by Nyerere included two main points:

1. "Principles of action can be set out, but the application of these principles must take into account the different geological conditions in different areas, and also the local variations in the basically similar traditional structures."[35]

2. Leadership and education – i.e. mentorship and coordination – are the most important resources the government can provide.

Neither one of the two guidelines saw genuine application in practice. So what happened? It is fair, in my opinion, to take the position that Nyerere was neither naive nor too romantic in his assessment of the Tanzanian reality, and that he pointed out many of the challenges that stood in the way of Ujamaa. Those challenges included the possibility that Tanzania may have already gone too far in the formulation of trends and classes unsupportive of a socialist vision. Moreover, Nyerere was writing and speaking to the people of Tanganyika/Tanzania directly. A leader in such a situation is wise to highlight, with pride, the positive aspects of the ways of his people. Building a coherent nation is not

---

(33) See Boesen 1977; Raikes 1975; and Croll 1979.

(34) Coulson (1984) argues that environmental problems did not hinder agricultural productivity during Ujamaa times, and that there were a number of cases that proved to be impressively productive during this time. Therefore Coulson affirms that the failure of Ujamaa was not due to technical-agronomical reasons but to failure in economic planning and management.

(35) Nyerere 1968, 121.

an isolated task from national development, and in that – building a nation – Nyerere's leadership has succeeded by most measures.[36] Being a national leader is not the same as being a leader or member of a small ideological group bound only by rationalistic reasoning. The walk is different, and the talk is different. Nyerere was evidently one of the rare post-colonial national leaders who walked their talk.

> "However, Nyerere had his limitation and that is the limitation of the visionary pioneer who finds himself in the position of the practitioner. For instance, when socialism was applied in the USSR, it was backed by more than a century of deep philosophical theorization. The African dilemma is that either it builds its practical on the Western theoretical, or it does them both simultaneously." (Hashim 2013)[37]

Without the need for going any further into the arguments presented above, let us concentrate on the indicators of inconsistency between planning and management throughout the Ujamaa scheme. De Vries (1978) talked about the role of extension workers in Ujamaa – public service, state-employed staff who were given the task of providing advice and technical support to the villagers – indicating that statistics show that the extension workers were in much more contact with the kulaks than with the rest of the villagers. The reasons were: (a) extension workers are themselves a privileged elite class who prefer to work with their own to sustain their status, and (b) most of the recommendations by the workers were readily suitable for the kulaks' capacity, which made the kulaks a better audience. This second reason shows that the management solutions for the villagers' problems were not realistic to the majority, which is an indicator of disconnect in a plan that claimed to be attentive to the sensitivities of the Tanzanian rural context. De Vries referred to how, under a system inherited from colonial rule (i.e. the system of extension workers), extension programs could not serve the revolutionary agenda of Ujamaa, which was based on a philosophy that doubted the colonial approach to development in the African context. An institutional change was needed to address this problem.

Besides the lack of evaluation research from field trials – an important adaptive management tool – and poor financial monitoring, mentioned by Freyhold (1979), she also pointed to the problem that the idea of Ujamaa was not an intellectual product of the ruling party TANU, but rather of a certain individual, namely Nyerere. The Ujamaa

---

(36)    See James 1977; Shivji 2012a and 1995.
(37)    M. Jalal Hashim (2013), in a personal e-mail exchange with the author.

philosophy did not show indicators of being deeply entrenched in the party ideology, especially its other senior members. Shivji makes the same observation: "It is revealing of Nyerere's political style and practice that there was no one in his party or the state to defend his ideology [after he resigned from leadership... although he remained widely genuinely respected]." (2012a, 112). According to this argument, the disconnect between planning and management was bound to happen when TANU did not found a system of training for its leading cadres to fully understand Ujamaa as "an attitude of mind," as Nyerere described it. Being the upper management team, their understanding the strategic plan and goals was crucial. Another problem that was pointed out by both Freyhold and Raikes (1975) is that not enough measures were taken by the government to monitor its staff in the villages. Many of the bureaucrats abused their power and the villagers could not trust them or report them.

Kjekshus (1977) brought attention to the fact that even when most critics of Ujamaa argued against it, they still agreed with its main thesis: that getting the "studded individual homesteads" of the rural population of Tanzania to resettle into collective settlements is a good thing. "Thus the critics of the scheme singled out the implementers and the implementation for censure while regarding the villagization plan as essentially sound."[38] From a sheer economic and administrative perspective, Ujamaa could not easily be argued against at the planning level. The criticism of the implementers and implementation was sometimes well-founded, however. For example, Kjekshus notes that, "The president's Ujamaa blueprints were late in gaining concrete formulations beyond the level of broad generalities. They were given the status of urgency and flawlessness through the 2nd Five Year Plan.... No comprehensive legislation dealing with the villages was forthcoming until 1975."

In the field of development, and especially on the executive side, connection and consistency between planning and management, as a dynamic relationship, is a requirement. On the one hand comprehensive and coherent understanding of plans is the most important information for the management team to possess. On the other hand, evaluation of the experience of the structures and processes of management feedback to planning so it adapts by means of reality checks. A perfect plan is not a requirement to achieve success in reality, and neither is a perfectly

---

(38)  Kjekshus 1977, 275.

coherent management system, but sufficient connection and consistency between the two is a requirement. Similar rural development schemes in other 'socialist' regimes around the world, around the same time as Ujamaa's, achieved more progress towards their self-ascribed goals than Ujamaa did. Whether in the collectivization of the USSR and China, or the agricultural development scheme in Cuba, productivity increased and the change in power relations delivered some positive results for the interest of women, for instance, to a greater degree than happened with Ujamaa.[39] In our claim, with more comparative investigation it could be found that one decisive factor that made those other collectivization schemes achieve more than what Ujamaa had was the relatively more apparent connection and consistency between planning and management. None of those schemes achieved everything that was set in their original plans, and neither were any able to continue without changing that original plan at certain clashes with reality during implementation, but different levels of success were still measurable.

The author is among those who still have admiration for the Ujamaa vision, and think that it still has a chance of 'critical resurrection' (i.e. being able to contribute to 'new and improved' rural development schemes), not only in Tanzania but in other parts of Africa as well. Indeed, Nyerere is among the big visionaries that have come out of post-colonial societies, "and like all visionaries, their legacy lies in their vision more than in their practice. While appropriating their vision, we need to examine their practice critically to draw lessons. Vision inspires, practice teaches." (Shivji 2008, xv-xvi)

---

(39)  Notwithstanding the overall critique any of those other experiences – villagization in the USSR, China and Cuba – may have received.

# Last Remarks

*"The movement of the Tao by contraries proceeds*
*And weakness marks the course of Tao's mighty deeds..."*

-Tao Te Ching[1]

One of the conclusions we can all draw, at the end of this book, is that transforming societies through technological change always takes place in sociopolitical contexts, beside economic ones. As said earlier, in chapter one, and according to Karl Polanyi, the economy itself "consists of technology employed within a context of institutions. This context is one of dynamic interaction. Institutions mould technology and technology moulds institutions."[2]

Yet, technology means more than that to us. Technology is a part of our evolution; it grew with us and contributed to our growth. At the cognitive and personal levels – which are also not isolated from sociopolitical and economic contexts – we are all encompassed by one big technosphere that moulds our ideas, behaviours and aspirations; all the more reason why we need to make a conscious effort to understand it and include it in how we approach our present, future, and the road traveled between them.

A quote that is attributed to Carl Sagan captures a global human dilemma: "We've arranged a global civilization in which most crucial elements profoundly depend on science and technology. We have

---

(1)     From the book of the Tao, *Tao Te Ching*, ancient Chinese classic text. According to tradition it was written around the 6th century BC by the sage Lao Tzu.

(2)     Stanfield 1990, 203-4.

also arranged things so that almost no one understands science and technology. This is a prescription for disaster. We might get away with it for a while, but sooner or later this combustible mixture of ignorance and power is going to blow up in our faces." It is a serious problem indeed, and it is catching up with us. This dilemma is further amplified in the context of developing societies that exist under conditions of technological dependency. If industrialized societies are at least still capable of relying on the current division of labour that accommodates the existence of small-but-sufficient groups of technical and scientific experts, and their hosting institutions, that have been collectively charged with managing the critical aspects of their societies' technosocial systems (i.e. the fifth estate) and the evolution of such systems, developing societies are not even there yet. Despite their own problems with how technology is institutionalized in their lives, industrialized societies as a whole are capable, to a visible extent, of creating and regulating most of their important technologies to render results that reflect in the modern qualities of life their people experience (albeit not yet equally, or fairly, among individuals, families and marginalized communities). Developing societies are yet to reach that point; and they ought to, at least. Some of us would say they must.

Pursuing technological autonomy should be a major goal for developing societies. It should be enshrined in the psyches of their peoples by the various media available, to have a collective appreciation of what is at stake. It should be seriously envisioned, planned for, and sought with concrete implementation policies and actions. That is because technological autonomy is tightly connected to the possibilities of genuine and sustainable development for all modern societies. Without increasing their technological capabilities, and localizing their technological processes and outputs, developing societies will remain on the margins of global affairs, now and in the future, with dire consequences for most aspects of their well-being (collectively and individually).

There are possibilities in crisis. History has taught us so. We currently have the element of crisis in abundance, but we also have accumulated trials, lessons, expertise, tools, and recognized agents of change. The possibilities are there, and the ingredients as well. Some developing societies have already demonstrated that, to a large degree, in the last few decades. Work is required, guided by understanding and energized by commitment. Technological autonomy is not likely to happen without conscious vision and planning. It has to be intentional and enduring over time and across multiple actors and institutions, and it has to be orientated toward material results. That is praxis.

One element that is not on our side, for sure, is time. The longer time elapses without qualitative change taking place, the slimmer the possibilities get. There might come a point in time when the gap of technological divide, between the technologically dependent and the technologically advanced, becomes too wide for bridging. This should be kept in mind.

Yet, technological autonomy is not an end goal in itself. A process toward autonomy needs to be oriented toward the immense possibilities of what is beyond reaching autonomy. It should be oriented towards liberation (as defined earlier in the book) and the various potentials of enjoying and advancing conditions of liberated societies and a liberated humanity at large. There is also the potential of not only 'catching up' in the technological march forward, but also becoming authentic contributors, in the foreseeable future, to the frontiers of that march. When we imagine such potential, we can imagine that our knowledge of the universe shall advance, and our technological feats shall advance with it, to take us as far as we can contemplate. Just because developing societies are currently faced with quite basic and immediate challenges, at the technological development level, it does not mean they cannot marry pragmatic focus on the now with unbound human ambition.

# References

Adeel, Zafar, Brigitte Schuster and Harriet Bigas (eds.) 2008. *What Makes Traditional Technologies Tick? A review of traditional approaches for water management in drylands.* Hamilton, Ontario: United Nations University, UNU-INWEH.

Adesina, Akinwumi A. and Jojo Baido-Forson. 1995. "Farmers' perceptions and adoption of new agricultural technology: evidence from analysis in Burkina Faso and Guinea, West Africa." *Agricultural Economics,* 13: 1-9.

Al-Ghafri, Abdullah S. 2008. "Traditional Water Distribution in Aflaj Irrigation Systems: Case Study of Oman" in Zafar Adeel, Brigitte Schuster and Harriet Bigas (eds.) *What makes traditional technology tick? A review of traditional approaches for water management in drylands.* Hamilton, Ontario: The United Nations University (UNU-INWEH). Chapter 8: 74-85.

Alter, Kim. 2007. *Social Enterprise Typology.* Report of Virtue Ventures LLC. Updated November 27.

An-Na'im, Abdullahi A. 2008. *Islam and the secular state: Negotiating the future of Shari'a.* Harvard University Press.

An-Na'im, Abdullahi A. (ed). 1995. *Human Rights in Cross-cultural Perspectives. A quest for consensus.* Philadelphia: Penn Press.

Ardant, Gabriel. 1963. "A Plan for Full Employment in the Developing Countries." *International Labour Review.*

Arthur, W. Brian. 1989. "Competing Technologies, Increasing Returns, and Lock-in by historical events." *The Economic Journal,* 99(394): 116-131.

Aunger, Robert. 2010. "Types of Technology." *Technological Forecasting & Social Change,* 77: 762-82.

Ash, Robert, Joseph Linn and C. J. Wu. 2006. "The Economic Legacy of the KMT" in Dafydd Fell, Henning Klöter & Bi-yu Chang's (eds.) *What has Changed? Taiwan Before and After the Change in Ruling Parties.* Harraswitz Verlag. Wiesbaden. Chapter 5, pp. 83 – 106.

Aubert, Jean-Eric. 2005. "Promoting Innovation in Developing Countries: A Conceptual Framework." Policy Research Working Paper; No. 3554. World Bank, Washington, DC.

Assefa, G. and B. Frostell. 2005. "Technology Assessment in the Journey to Sustainable Development", in Gedeon M. Mudacumura, Desta Mebratu & M. Shamsul Haque (eds.) *Sustainable Development Policy and Administration.* Taylor & Francis.

Baldacchino, Godfrey. 1990. "A War of position: Ideas on a strategy for worker cooperative development." *Economic and Industrial Democracy,* 11: 463-482.

Basurto, Xavier and Ostrom, Elinor. 2009. "The Core Challenges of Moving Beyond Garrett Hardin." *Journal of Natural Resources Policy Research,* 1(3): 255-59.

Bhattacharyya, Jnanabrata. 2004. "Theorizing community development." *Journal of Community Development Society,* 34(2): 5-34.

Binswanger, Hans. 1986. "Agricultural Mechanization: A comparative historical perspective." *World Bank Research Observer,* 1(1): 27-56.

Borgmann, Albert. 2010. "Reality and Technology." *Cambridge Journal of Economics,* 34: 27-35.

Biersteker, Thomas J. 1980. "Self-Reliance in Theory and Practice in Tanzanian Trade Relations." *International Organization,* 34(2): 229-264.

Biko, Stephen. 1978. *I Write What I Like: Selected Writings.* University of Chicago Press (2002).

Boesen, Jannik, Bridgit Storgaard Madsen and Tony Moody. 1977. *Ujamaa—Socialism From Above.* New York, NY: Africana Publishing Company.

Brint, Steven, Kristopher Proctor, Robert A. Hanneman, Kerry Mulligan, Matthew B. Rotondi, and Scott P. Murphy (2011). "Who are the early adopters of new academic fields? Comparing four perspectives on the institutionalization of degree granting programs in US four-year colleges and Universities, 1970–2005." *Higher Education,* 61(5): pp 563-585.

Bryson, J. M. 2004. "Strategic Planning." *International Encyclopaedia of the Social & Behavioural Sciences,* pp. 15145-15151. Amsterdam: Elsevier.

Cabral, Amilcar. 1966. *The Weapon of Theory.* (Address delivered to the first Tricontinental Conference of the Peoples of Asia, Africa and Latin America held in Havana in January).

Canadian Clean Technology Industry Report: 2013 Prospectus (2012).

Commission for Science & Technology (COSTECH), Tanzania. nd. "Directorate, Centre for Development and Transfer of Technology." Accessed August 15, 2013. http://www.costech.or.tz/?page_id=1657

Coulson, Andrew. 1982. *Tanzania: a Political Economy.* Oxford: Oxford University Press. (2nd edition 2013).

Crabitès, Pierre. 1912. "Egypt, the Sudan and the Nile." *Foreign Affairs* 6(2): 320-328.

Croll, Elizabeth J. 1979. *Socialist Development Experience: Women in rural production and reproduction in the Soviet Union, China, Cuba and Tanzania.* Discussion paper no.143 for the Institute of Development Studies at the University of Sussex, Brighton, England.

Dacanay, Marie Lisa M. 2012. *Social enterprises and the poor: Enhancing social entrepreneurship and stakeholder theory.* Dissertation: Doctoral School of Organization and Management Studies, Copenhagen Business School, Denmark.

Dalton, George. 1990. "Writings that clarify theoretical disputes over Karl Polanyi's work" in Kari Polanyi-Levitt's (ed.) *The life and work of Karl Polanyi.* Montreal: Black Rose Books. Chapter 18: 161-170.

Dandekar, H. C. 2004. "Rural Planning: General." *International Encyclopaedia of the Social & Behavioural Sciences,* pp. 13425-13429. Amsterdam: Elsevier.

Davies, R. W. 1980. *The Socialist Offensive: The Collectivization of Soviet agriculture 1929-1930.* London and Basingstoke: MacMillan Press Ltd.

De Araujo, Manuel G. M. 1985. "Communal villages and the distribution of the rural population in the People's Republic of Mozambique," in John I. Clarke, Mustafa Khojali & Leszek A. Kosinski's (eds.) *Population and Development Projects in Africa.* Cambridge: Cambridge University Press.

de Mabior, John Garang. 1981. *Identifying, Selecting, and Implementing Rural Development Strategies for Socio-economic Development in the Jonglei Projects Area, Southern Region, Sudan.* PhD Dissertation, Iowa State University, USA.

de Mabior, John Garang. 1992. *"The Call for Democracy in Sudan".* Edited by Mansour Khalid

Dercon, Stefan, and Luc Christaiensen (2011). "Consumption risk, technology adoption and poverty traps: Evidence from Ethiopia." *Journal of Development Economics,* 96: 159-173.

De Vries, James. 1978. "Agricultural Extension and Development— Ujamaa villages and the problems of institutional change." *Community Development Journal,* 13(1): 11-19.

Dengu, Ebbie, El-Garrai, Omer & Abdalla, Asim H. 2006 (May). *Evaluation of The Food Security Project: Re-Establishing Food Self-Reliance Amongst Drought Affected People of North Darfur.* Khartoum: Practical Action Sudan.

Department of Trade and Industry, Great Britain. 2002. *Social Enterprise: a strategy for success.*

Desai, Meghnad, Sakiko Fukuda-Parr, Claes Hohansson & Fransisco Sagasti. 2002. "Measuring the Technology Achievement of Nations and the Capacity to Participate in the Network Age." *Journal of Human Development,* 3(1): 95-122.

Devereux, Stephen and Edwards, Stephen. 2004. "Climate Change and Food Security." *Institute of Development Studies (IDS) Bulletin,* Vol. 35(3): 22-30.

Diamond, Larry. 2010. "Liberation Technology." *Journal of Democracy,* Vol. 21(3): 69-83.

Diop, Cheikh Anta. 1988. *Precolonial Black Africa.* translated by Harold Salemson (from French). Chicago Review Press.

Diop, Cheikh Anta. 1974. *The African Origin of Civilization: Myth or Reality?* Chicago Review Press (1989).

Dosi, Giovanni. 1982. "Technological paradigms and technological trajectories: A suggested interpretation of the determinants and directions of technological change." *Research Policy,* 11 (1982): 147-162.

Dutta, Soumitra & Lanvin, Bruno (eds.) 2013. *Global Development Index Report: The Local Dynamics of Innovation.* Geneva: World Intellectual Property Organization (WIPO), and New Delhi: Confederation of Indian Industry (CII).

*Dwivedi, Ranjit. 1998. "Resisting Dams and 'Development': Contemporary Significance of the Campaign Against the Narmada Projects in India." European Journal of Development Research, 10(2): 135-183.*

Eisler, Riane. 2002. "The Dynamics of Cultural and Technological Evolution: Domination Versus Partnership." *World Futures,* 58(2&3): 159-174.

El-Khalifa, Abubakr. 2007. "Is the Hidden Intent of Merowe Dam to Rid Egypt of the High Dam's Problems?" (Arabic). Sudanile, March.

Encyclopedia of Modern Asia. 2006. "Agricultural Collectivization —China." Accessed January 16, 2010: http://www.bookrags.com/research/agricultural-collectivization-china-ema-01/

Ferguson, Niall. 2011. *Civilization: the West and the Rest.* The Penguin Press HC

Fidiel, Mohamed M. 2006. *Technology and Gender Issues: the experience of Practical Action Sudan (Arabic).* Practical Action Publications.

Fidiel, Mohamed M. 2005. *Building Small Scale Water Harvesting Dams: The experience of intermediate technology development group, North Darfur State - Western Sudan.* Accessed September 10, 2009: http://practicalaction.org/sudan/docs/region_sudan/water-harvesting.pdf

Forje, John W (1989). *Science and Technology in Africa.* Longman Group UK Limited.

Franklin, Ursula. 1989. *The Real World of Technology.* CBC Massey lectures, Tuesday, November 7th. Accessed August 25, 2013 from CBC Massey Lectures Archives: http://www.cbc.ca/ideas/massey-archives/1989/11/07/1989-massey-lectures-the-real-world-of-technology/

Franssen, Maarten, Gert-Jon Lokhorst & Ibo van de Poel. 2013. "Philosophy of Technology." *The Stanford Encyclopedia of Philosophy (Fall 2015 Edition),* Edward N. Zalta (ed.), Accessed November 2015: http://plato.stanford.edu/archives/fall2015/entries/technology/

*Freire, Paulo. 1984. Pedagogy of the Oppressed. Translated by Bergman Ramos. New York: Continuum.*

Freyhold, Michaela Von. 1979. *Ujamaa Villages in Tanzania: Analysis of a Social Experiment.* New York, NY: Monthly Review Press.

Galtung, Johan. 1979. *Development, Environment and Technology: towards a technology for self-reliance.* New York: United Nations.

Gambino, Christine P., Edward N. Trevelyan and John Thomas Fitzwater. 2014. "The Foreign-Born Population From Africa: 2008–2012" *American Community Survey Briefs.* US Census Bureau: ACSBR/12-16.

Gamser, Matthews S. 1988. "Innovation, Technical Assistance and Development: The Importance of Technology Users." *World Development,* 16(6): 711-721.

Gilbert, Jess. 2009. "Democratizing states and the use of history." *Rural Sociology,* 74(1): 3-24.

Gill, Graeme. 2003. *The Nature and Development of the Modern State.* New York: Palgrave Macmillan.

Granados, Maria L., Vlatka Hlupic, Elayne Coakes and Souad Mohamed. 2011. "Social enterprise and social entrepreneurship research and theory: A bibliometric analysis from 1991 to 2010." *Social Enterprise Journal,* 7(3): 198-218.

Greenstone, Michael and Adam Looney. 2010. "Ten Economic Facts about Immigration." Report by the Hamilton Project, Brookings Institution, USA. Accessed November 2015: http://www.brookings. edu/~/media/research/files/reports/2010/9/immigration-greenstone-looney/09_immigration.pdf

Gulrajani, Mohini. 2006. "Technological Capabilities in Industrial Clusters: A Case Study of Textile Cluster in Northern India." *Science, Technology & Society,* 11(1): 149-90.

Haider, Muhiuddin, and Gary L. Kreps. 2010. "Forty Years of Diffusion of Innovations: Utility and Value in Public Health." *Journal of Health Communication,* Vol. 9(1): 3-11.

Harriss, John. 2003. "Institutions, Politics and Culture: a Polanyian Perspective on Economic Change." *International Review of Sociology,* 13(2): 343-355.

Harvey, Mark, Ronnie Ramlogan and Sally Randles (eds.) 2007. *Karl Polanyi: New perspectives on the place of the economy in society.* Manchester: Manchester University Press.

Haug, David M. 1992. "The international transfer of technology: Lessons that east Europe can learn from the failed third world experience." *Harvard Journal of Law & Technology,* 5 (Spring Issue): 209-240

Hawken, Paul, Amory Lovins & L. Hunter Lovins. 2000. *Natural Capitalism: Creating the Next Industrial Revolution.* US Green Building Council.

Hewlett, Sylvia Ann, Melinda Marshall & Laura Sherbin (2013). "How Diversity Can Drive Innovation." *Harvard Business Review,* December Issue.

Hill, Ronald Paul &Kanwalroop Kathy Dhanda. 2003. "Technological Achievement and Human Development: A View from the United Nations Development Program." *Human Rights Quarterly,* 25(4): 1020-1034

Hobson, John M. 2004. *The Eastern Origins of Western Civilisation.* Cambridge: Cambridge University Press.

Hodgson, Geoffrey. 2007. "The enforcement of contracts and property rights: Constitutive versus epiphenomenal conceptions of law" in Mark Harvey, Sally Randles and Ronnie Ramolgan (eds.) *Karl Polanyi: New perspectives on the place of the economy in society.* Manchester: Manchester University Press. Chapter 4: 58-77.

Hodgson, Geoffrey. 2004. "Institutional Economic Thought" *International Encyclopedia of the Social & Behavioral Sciences,* Pages 7543-7550. Amsterdam: Elsevier.

Hopkins, Terence K. 1957. "Sociology and the Substantive View of the Economy" in Polanyi, K, Conrad M. Arensberg and Harry W. Pearson's (eds.) *Trade and market in early empires: Economies in history and theory.* Glencoe, IL: The Free Press. Chapter 14: 270-307.

Howes, Michael and Chambers, Robert. 1979. "Indigenous technical knowledge: Analysis, Implications and Issues." *Institute of Development Studies (IDS) Bulletin,* 10.2

Huh, Y.E. and S.H Kim. 2008. "Do early adopters upgrade early? Role of post-adoption behavior in the purchase of next-generation products." Journal of Business Research, 61(1): 40-46.

Hyden, Goran. 2006. "Introduction and Overview to the Special Issue on Africa's Moral and Affective Economy." *African Studies Quarterly,* 9(1 & 2).

Hyden, Goran. 1980. *Beyond Ujamaa in Tanzania: Underdevelopment and an uncaptured peasantry.* Berkeley and Los Angeles, CA: University of California Press.

Hyman, Eric L. 1987. "The Identification of Appropriate Technologies for Rural Development." *Impact Assessment,* 4(5): 35-55.

Industrial Technology Research Institute (ITRI), Taiwan. nd.. "About Us." Accessed March 8, 2016: https://www.itri.org.tw/eng/Content/Messagess/contents.aspx?SiteID=1&MmmID=617731521661672477

Intergovernmental Panel on Climate Change – IPCC. 2007. *Climate Change 2007: Impacts, Adaptation and Vulnerability.* Working Group II contribution to the IPCC Fourth Assessment Report, Summary for Policymakers. Pg. 3.

International Federation of Agricultural Producers. 2005. *Good Practices in Agricultural Water Management: Case Studies from Farmers Worldwide.* Background Paper Number. 3. United Nations Department of Economic and Social Affairs, Commission on Sustainable Development, thirteenth session, 11-22 April.

James, C.L.R. 1977. *Nkrumah and the Ghana revolution.* London: Allison and Busby. Cited in Issa G. Shivji's "Nationalism and pan-

Africanism: decisive moments in Nyerere's intellectual and political thought." *Review of African Political Economy*, 39(131): 103–116.

Japan Automobile Manufacturers Association (JAMA). nd. "Towards Industrialization (1935-1945)". Accessed August 20, 2013: http://njkk.com/about/industry2.htm

Jones, Monty. 2009. "Key Challenges for Technology Development and Agricultural Research in Africa." *IDS Bulletin*, Vol. 34(2): 46-51.

Jossa, Bruno. 2005. "Marx, Marxism and the Cooperative Movement." *Cambridge Journal of Economics*, 29(1): 3-18.

Jossa, Bruno. 2009. "Gramsci and the labor-managed farm." *Review of Radical Political Economics*, 41(1): 5-22.

Kandji, Serigne Tacko, Louis Verchot & Jens Mackensen. 2006. "Climate Change and Variability in the Sahel Region: Impacts and Adaptation Strategies in the Agricultural Sector." Report by World Agroforestry Centre (ICRAF) and the United Nations Environment Programme (UNEP).

Kavanaugh, Andrea. 1998. *The Social Control of Technology in North Africa: Information in the Global Economy*. Westport, Connecticut, Praeger, p. xii preface.

Kerr, John. 2002. "Watershed Development, Environmental Services, and Poverty Alleviation in India." *World Development*, 30(8): 1387-1400.

Khagram, Sanjeev. 2004. *Dams and Development: Transnational Struggles for Water and Power*. Ithaga and London: Cornell University Press.

Kim, Linsu & Nelson, Richard R. (eds.) 2000. *Technology, Learning and Innovation: Experiences of Newly Industrializing Economies*. New York: Cambridge University Press.

Kjekshus, Helge. 1977. "The Tanzanian Villagization Policy: Implementational Lessons and Ecological Dimensions." *Canadian Journal of African Studies*, 11(2): 269-282.

Kroszner, Randall. 1987. "Technology and Control of Labor" A review of *Forces of Production: a Social History of Industrial Automation* by David F. Noble. *Critical Review*, 1(2): 6-16.

Lall, Sanjaya. 1992. "Technological capabilities and industrialization." *World Development*, 20(2): 165-186.

Lekoko, Rebecca N. and Semali, Ladislaus. 2012. *Cases on developing countries and ICT integration: Rural community development*. IGI Global.

Loxley, John and John S. Saul. 1975. "Multinationals, workers and the parastatals in Tanzania." *Review of African Political Economy,* 2: 54-88.

Lucena, Juan, Jen Schneider and Jon A. Leydens. 2010. *Engineering and sustainable community development.* Morgan & Claypool Publishers.

Lundvall, Bengt-Åke (ed.). 1992. *National Innovation Systems: Towards a Theory of Innovation and Interactive Learning.* London: Pinter.

Maharajh, Rasigan, Mrio Scerri and Malean Sibanda. 2013 (March). *A review of the National Innovation System of the United Republic of Tanzania: External Review Report.*

Mair, Johanna, and Oliver Schoen. 2007. "Successful social entrepreneurial business models in the context of developing economies." *International Journal of Emerging Markets.* Vol 2(1): 54-68.

Martinez-Torres, Maria Elana, Peter M. Rosset. 2010. "La Via Campesina: The birth and evolution of transnational social movement." *Journal of Peasant Studies,* 37(1): 149-175.

Marx, Karl. 1977. A Contribution to the Critique of Political Economy. Moscow: Progress Publishers. (original manuscript published 1859)

Marx, Karl. 1887. Capital, Volume I. Moscow: Progress Publishers. (original manuscript published 1867)

Mazrui, Ali Al'Amin. 1986. "Tools of Exploitation." Part 4 of *The Africans: A Triple Heritage,* a BBC documentary series.

Mazzucato, Mariana. 2013. The Entrepreneurial State: Debunking public vs. private sector myths. New York: Anthem Press.

Menard C. and and M. M. Shirley (eds.) 2005. *Handbook of New Institutional Economics.* The Netherlands: Springer.

Mintesinot B., W Kifle & T Leulseged. 2004. "Fighting famine and poverty through water harvesting in Northern Ethiopia", in *Comprehensive Assessment Bright Spots Final Project,* compiled by A. Noble. Colombo, Sri Lanka: International Water Management Institute.

Miraftab, Faranak. 2004. "Public-Private partnerships: The Trojan horse of neoliberal development?" *Journal of Planning Education and Research,* 24: 59-101.

Monden, Yasuhiro. 1993. *Toyota Production System: An Integrated Approach to Just-In-Time (2nd ed).* Industrial Engineering and Management Press, Institute of Industrial Engineers, Norcross, Georgia.

Morehouse, Ward. 1979. "Science, Technology, Autonomy, and Dependence: A Framework for International Debate." *Alternatives*, IV:387-412

Morgan, M. Granger. 2010. "Technology and Policy", chapter 19 in Domenico Grasso and Melody Brown Burkins (eds) *Holistic Engineering Education: Beyond Technology*, pages 271-281, New York: Springer.

Mumford, Lewis. 1967. *The Myth of the Machine: Technics and Human Development (Vol. 1)*. Mariner Books.

Mumford, Lewis.1970. *The Myth of the Machine: The Pentagon of Power (Vol. 2)*. Harcour, Brace Jovanovich.

Nasir, Anthony, Tariq Mahmood Ali, Sheikh Shahdin & Tariq Ur Rahman. 2011. "Technology achievement index 2009: ranking and comparative study of nations." *Scientometrics*, 87(1):41–62

Noble, David. 1984. *Forces of Production: A Social History of Industrial Automation*. New York: Knopf.

North, Douglas. 1990. *Institutions, Institutional Change and Economic Performance*. Cambridge: Cambridge University Press.

Nyerere, Julius. 2011. *Freedom, Non-alignment and South-South Cooperation (A selection from speeches 1974-1999)*. Dar es Salaam: Oxford University Press.

Nyerere, Julius. 1998. Address by Chairman of the South Centre, to opening of conference on 'Governance in Africa. Addis Ababa, March 2.

Nyerere, Julius. 1968. *Ujamaa—Essays on Socialism*. Dar es Salaam: Oxford University Press.

O'Brien, Robert and Marc Williams. 2004. *Global Political Economy: Evolution and Dynamics* (2nd ed). New York: Palgrave Macmillan.

OECD. 1993. *Effective Technology Transfer and Co-operation for Environmentally Sustainable Strategies: Draft Common Reference Paper*. DCD/DAC/ENV (93)4.

OECD. 2010a. *Comparing nuclear accident risks with those from other energy sources*. Report No. 6861 by the OECD Nuclear Energy Agency (NEA).

OECD. 2010b. *Transition to a Low-carbon Economy: Public Goals and Corporate Practices*.

Orrnert, Anna. 2006. "Missing pieces: An overview of the institutional puzzle." *Public Administration and Development*, 26: 449-55.

Ostrom, Elinor. 2005. "Doing Institutional Analysis: Digging Deeper than Markets and Hierarchies" in C. Me´nard and M. M. Shirley (eds.) *Handbook of New Institutional Economics.* The Netherlands: Springer. Chapter 30: 819–848.

Page, John. 2016. "Industry in Tanzania: Performance, prospects, and public policy." Working paper 2016/5 of the United Nations University - World Institute for Development Economic Research (UNU-WIDER).

Pande, Rohini. 2007. "Large Dams in India," in Kaushik Basu (ed.) *The Oxford Companion to Economics in India.* Oxford University Press.

Patel, C. C. 1995. "The Sardar Sarovar Project: A Victim of Time." Chapter 3 In *William F. Fisher (ed.) Toward Sustainable Development: Struggling Over India's Narmada River. London: M. E. Sharpe. 71-112.*

Patterson, R. and J. Bozeman. 1999. "Comparativist Study of State and Promotion of Science and Technology. Cases: Botswana and Singapore", in R. Patterson, (ed.) *Science and Technology and Southern African and East and South Asia,* 116-42. Brill: Leiden.

Parry, Martin, Cynthia Rosenzweig, Matthew Livermor. 2005. "Climate change, global food supply and risk of hunger." *Philosophical Transactions: Biological Sciences,* 360(1463): 2125-2138.

Pieterse, Heinie. 2001. Telecommunications Technology Transfer/ Diffusion Model into Rural South Africa. Master of Engineering Dissertation Summary, University of Pretoria. Accessed August 11, 2012: http://upetd.up.ac.za/thesis/available/etd-05102002-141643/unrestricted/00front.pdf

Pink, Joanna (ed). 2009. *Muslim Societies in the Age of Mass Consumption: Politics, Culture and Identity between the local and the Global.* Newcastle: Cambridge Scholars Publishing.

Polanyi, Karl, Conrad M. Arensberg and Harry W. Pearson's (eds.) 1957. *Trade and market in early empires: Economies in history and theory.* Glencoe, IL: The Free Press.

Polanyi, Karl. 1957. "The economy as instituted process" in Polanyi, K, Conrad M. Arensberg and Harry W. Pearson (eds.) *Trade and market in early empires: Economies in history and theory.* Glencoe, IL: The Free Press. Chapter 13: 243-70.

Polanyi, Karl. 1968. *Primitive, Archaic and Modern Economies: Essays of Karl Polanyi,* edited by George Dalton. Garden City, NY: Anchor Books.

Polanyi, Karl. 1944. *The Great Transformation.* Boston: Beacon Press

Practical Action. 2002. *Water Harvesting in Sudan.* Technical Brief, Warwickshire: The Schumacher Centre for Technology and Development.

Prah, Kwesi Kwaa. 2009. "The Burden of English in Africa: from Colonialism to Neo-colonialism." Keynote Address presented to the Department of English: *5th International Conference on the theme: Mapping Africa in the English-Speaking World.* University of Botswana. 2nd – 4th June.

Raffensperger C. & Tickner, J. A. (eds.) 1999. *Protecting Public Health and the Environment: Implementing the Precautionary Principle.* Island Press, Washington, DC

Ram, Rahul N. 1995. "Benefits of the Sardar Sarovar Project: Are the Claims Reliable?" Chapter 5 In *William F. Fisher (ed.) Toward Sustainable Development: Struggling Over India's Narmada River. London: M. E. Sharpe. 113-34.*

Raikes, P. L. 1975. "Ujamaa and Rural Socialism." *Review of African Political Economy,* Issue 3: 33-52.

*Ramachandra, Komala. 2006. "Sardar Sarovar: An Experience Retained?" Harvard Human Rights Journal, (19): 275-81.*

Rauniyar, Ganesh P. and Frank M. Goode. 1992. "Technology Adoption on Small Farms." *World Development,* Vol. 20(2): 275-82.

Rensburg, Johann Van, Allda Veidsman and Micheal Jenkins. 2008. "From Technologists to Social Enterprise Developers: Our Journey as "ICT for Development" Practitioners in Southern Africa." *Information Technology for Development,* Vol. 14(1): 76-89.

Ridley-Duff, Rory and Mike Bull. 2011. *Understanding Social Enterprise: Theory and Practice.* SAGE Publications.

Ridley-Duff, Roley. 2009. "Co-operative social enterprises: company rules, access to finance and management practice." *Social Enterprise Journal,* 5(1): 50-68

Rockefeller Foundation (and Global Business Network). 2010 (May). *Scenarios for the Future of Technology and International Development.* Report.

Rodgers, Loren. 2008. "Hybrid Cooperatives: Challenges and Advantages." National Center for Employee Ownership, January 30.

Rodney, Walter. 1972a. *How Europe Underdeveloped Africa* (6th reprint, 1983). Dar-Es-Salam: Tanzania Publishing House (TPH).

Rodney, Walter. 1972b. "Tanzanian Ujamaa and Scientific Socialism." *African Review* [Dar es Salaam, Tanzania], 1(4): 61-76.

Rogers, Everett. 2003. *Diffusion of Innovations* (5th ed.) Toronto: Free Pres.

Rogers, Everett, Arvind Singhal, and Margaret M. Quinlan. 2009. "Diffusion of Innovations" in Stacks, D. W and Michael B. Swalon's (eds.) *An Integrated Approach to Communication Theory and Research* (2nd *ed.)* NY: Routlodge. Chapter 27, pp. 418-434.

Rosenberg, Nathan. 1982. *Inside the Black Box: Economy and Technology.* New York: Cambridge University Press.

Rosenberg, Nathan. 1972. "Factors affecting the diffusion of technology." *Explorations in Economics History,* Fall.

Roy, Tirthankar. 2002. "Acceptance of innovation in early twentieth-century Indian weaving." *Economic History Review,* LV(3): 507-532.

Ruiz, Neil G. 2014. "The Geography of Foreign Students in U.S. Higher Education: Origins and Destinations". Brookings Institution. Accessed November 2015: http://www.brookings.edu/research/ interactives/2014/geography-of-foreign-students#/M10420

Ruttan, Vernon W. 1997. "Induced Innovation, Evolutionary Theory and Path Dependence: Sources of Technical Change." *The Economic Journal,* 107 (444): 1520-1529

Saul, John S. 2012. "Tanzania fifty years on (1961–2011): Rethinking Ujamaa, Nyerere and Socialism in Africa." *Review of African Political Economy,* 39(131): 117-125.

Schneider, Leander. 2004. "Freedom and Unfreedom in Rural Development: Julius Nyerere, Ujamaa Vijijini, and Villagization." *Canadian Journal of African Studies,* 38(2): 344-392

Schumacher, E. F. 1973. *Small is Beautiful: A Study of Economics as if People Mattered* (2nd ed.) London: ABACUS.

Science, Technology and Innovation Policy Research Organization (STIPRO). 2010 (June). "The Utility Value of Research and Development (R&D): Where does Tanzania Stand?". Policy Brief, Vol. 1, Issue 1.

Scott, James C. 1999. *Seeing Like a State: How Certain Schemes to Improve the Human Condition Have Failed.* Yale University Press.

Seely, Bruce E. 2003. "Historical Patterns in the Scholarship of Technology Transfer." *Comparative Technology Transfer and Society,* 1(1): 7-48.

Sen, Amartya. 1999. *Development as Freedom.* Toronto: Random House.

Shah, Ashvin A. 1995. "A Technical Overview of the Flawed Sardar Sarovar Project and a Proposal for a Sustainable Alternative." Chapter 13 In *William F. Fisher (ed.) Toward Sustainable Development: Struggling Over India's Narmada River. London: M. E. Sharpe.* 319-67.

Shaw, K.E. 2002. "Education and Technological Capability Building in the Gulf." *International Journal of Technology Design and Education,* 12(1): 77-91.

Sheikheldin, Gussai H. 2017. "Social Enterprises as Agents of Technological Change: Case Studies from Tanzania." PhD diss. University of Guelph. http://hdl.handle.net/10214/10363

Sheikheldin, Gussai H. 2015a. "Ujamaa: Planning and Development Schemes in Africa, Tanzania as a Case Study". *Journal of Pan-African Studies*, 8(1): 78-96.

Sheikheldin, Gussai H. 2015b. "Telling Histories, Telling Stories." Text introduction to artistic exhibition *Local Heroes* at Oslo Kunstnerforbundet, Norway, by C. Fadlabi, April 30th to May 27th. http://fadlabihimself.com/local-heroes/

Sheikheldin, Gussai H. 2013. "Ethiopian Renaissance Dam: Friend or Foe?" *The Citizen (newspaper)*, Sudan, June 2nd.

Sheikheldin, Gussai, Gail Krantzberg and Karl Schaefer. 2010. "Science-seeking behaviour of Conservation Authorities in Ontario." *Environmental Management*, 45(5): 912-921.

Shivji, Issa. 2012a. "Nationalism and pan-Africanism: decisive moments in Nyerere's intellectual and political thought." *Review of African Political Economy*, 39(131): 103–116.

Shivji, Issa. 2012b. "Remembering Walter Rodney." *Monthly Review*, 64(7).

Shivji, Issa G. 2008. *Pan-Africanism or Pragmatism: Lessons of the Tanganyika-Zanzibar Union.* Dar es Salaam: Mkukuna Nyota Publishers.

Shivji, Issa. 1995. "The Rule of Law and Ujamaa in the Ideological Formation of Tanzania." *Social and Legal Studies*, 4: 147-174.

Siddig, Mohamed. 2004."Challenging climate change: Terracing technology in Darfur." *Sharing*, Practical Action Sudan Newsletter. April.

Simalenga, Timothy E. 1999. "The Animal Traction Network in East and Southern Africa" in Devlin, J. D and T. Zettel's (eds.) *Ecoagriculture: Initiatives in Eastern and Southern Africa.* Harare: Weaver Press. Chapter 27: 303-314.

*Singh, Satyajit. 1997. Taming the Waters: The Political Economy of Large Dams in India. Delhi: Oxford University Press.*

Spreckley, Freer. 1981. *Social Audit – A Management Tool for Co-operative Working.* Beechwood College

Stamp, Patricia. 1990. *Technology, Gender, and Power in Africa.* Ottawa: International Development Research Centre.

Stanfield, J. Ron. 1990. "Karl Polanyi and contemporary economic thought" in Kari Polanyi-Levitt's (ed.) *The life and work of Karl Polanyi*. Montreal: Black Rose Books. Chapter 22: 195-207.

Swain, Ashok. 1997. "Ethiopia, the Sudan and Egypt: The Nile River Dispute." *The Journal of Modern African Studies* 35(4): 675-694.

Taha, Mahmoud M. 1958. "Education: A Letter Address to Mr. Osman Mahjoub, Dean of the Institute of Bakht al-Rida." 24 Dec, Khartoum. Accessed January 11' 2016: http://www.alfikra.org/article_page_view_e.php?article_id=1155&page_id=1

Tandon, Yash. 2015. *Trade is War*. Dar es Salaam: Mkuki na Nyota Publishers.

Teodoro, Cristian, Alfred Wüest & Bernhard Wehrli. 2006 (March 23). *Independent Review of the Environmental Impact Assessment for the Merowe Dam Project (Nile River, Sudan)*. Switzerland: Eawag, the Swiss Federal Institute of Aquatic Science and Technology.

Tustian, R. E. 2004. "Administrative Organization of Planning." *International Encyclopaedia of the Social & Behavioural Sciences*, pp. 11469-11474. Amsterdam: Elsevier.

United Nations Conference on Trade and Development, Secretariat. 1981. *Planning the Technological Transformation of Developing Countries: Study, Part 49*.

United Nations Convention on Biological Diversity – UNCBD. 2007. *Traditional knowledge and the Convention on Biological Diversity*. brochure. Accessed August 18, 2010: http://www.cbd.int/doc/publications/8j-brochure-en.pdf

United Nations Development Programme. 2001. *Making New Technologies Work for Development*. (Summary report)

United Nations Environmental Programme. 2008. *Indigenous knowledge in Disaster Management in Africa*. Publication compiled and edited by Peter Mwaura

Verschuren, Piet J. M. 2003. "Case study as a research strategy: Some ambiguities and opportunities." *International Journal of Social Research Methodology*, 6(2), 121-139.

Visvanathan, S. 2004. "Technology Transfer." *International Encyclopedia of the Social & Behavioral Sciences*, Pages 15532-37. Amsterdam: Elsevier.

Voss, T. R. 2004. "Institutions." *International Encyclopedia of the Social & Behavioral Sciences*, Pages 7561-66. Amsterdam: Elsevier.

Wejnert, Barbara. 2002. "Integrating Models of Diffusion of Innovations: a Conceptual Framework." *Annual Review of Sociology*, 28: 297-326

Wittfogel, Karl. 1957. *Oriental Despotism: A Comparative Study of Total Power*. New Haven: Yale University Press.

Wolff, Peter. 1999. *Vietnam, the Incomplete Transformation*. Psychology Press.

Woodhouse, Edward, and Patton, Jason W. 2004. "Introduction: Design by Society: Science and Technology Studies and the Social Shaping of Design." Design Issues 20(3): 1-12.

World Bank. 1988. Parastatals in Tanzania: towards a reform program. Report No. 7100-TA. Country IV Department, Africa Region. July 27.

World Commission on Dams. 2000. *Dams and Development: A New Framework for Decision-making*. (official report).

Worthington, Steve, Frauke Mattison Thompson, and David B. Stewart. 2011. "Credit cards in a Chinese cultural context—The young, affluent Chinese as early adopters." *Journal of Retailing and Consumer Services*, 18(6): 534-541.

Zafarullah, H. And A. S Huque. 2005. "Understanding Development Governance: Concepts, Institutions and Processes." In Huque, A. S and H. Zafarullah (eds.) *International Development Governance*. New York: Dekker/CRC Press.